Maturin Murray Ballou

The new Eldorado

a summer journey to Alaska

Maturin Murray Ballou

The new Eldorado
a summer journey to Alaska

ISBN/EAN: 9783744745680

Printed in Europe, USA, Canada, Australia, Japan

Cover: Foto ©Andreas Hilbeck / pixelio.de

More available books at **www.hansebooks.com**

THE NEW ELDORADO

A SUMMER JOURNEY TO ALASKA

BY

MATURIN M. BALLOU

I pity the man who can travel from Dan to Beersheba, and cry: " 'T is all barren!" and so it is, and so is all the world to him who will not cultivate the fruits it offers. — STERNE.

THIRD EDITION.

BOSTON AND NEW YORK
HOUGHTON, MIFFLIN AND COMPANY
The Riverside Press, Cambridge
1890

PREFACE.

THE Spaniards of old had a proverb signifying that he who would bring home the wealth of the Indies must carry the wealth of the Indies with him. If we would benefit by travel we must take with us an ample store of appreciative intelligence. Nature, like lovely womanhood, only reveals herself to him who humbly and diligently seeks her. As Sir Richard Steele said of a certain noble lady: " To love her is a liberal education." Keen observation is as necessary to the traveler who would improve by his vocation as are wings to an albatross. The trained and appreciative eye is like the object-glass of the photographic machine, nothing is so seemingly insignificant as to escape it. Careless, half-educated persons are sent upon their travels in order, it is said, that they may " learn." Such individuals had best first learn to travel. Those who improve the modern facilities for seeing the world acquire an inexhaustible wealth of information, and a delightful mental resort of which nothing can deprive them. The power of vision is thus enlarged, many occurrences which have heretofore proved daily mysteries become clear, prejudices are annihilated, and the judgment broadened. Above all, let us first become familiar with the important features of our own beautiful and widespread land before we seek foreign shores, especially as we have on this continent so much of unequaled grandeur and unique phenomena to satisfy and to attract us. It seems to the undersigned that perhaps this volume will have a tendency to lead the reader to such conclusion, and certainly this is its primary object.

M. M. B.

CONTENTS.

CHAPTER I.

Itinerary. — St. Paul. — The Northern Pacific Railroad. — Progress. — Luxurious Traveling. — Riding on a Locomotive. — Night Experiences. — Prairie Scenes. — Immense Grain-Fields. — The Badlands. — Climbing the Rocky Mountains. — Cinnabar. — The Yellowstone Park. — An Accumulation of Wonders. — The Famous Hot Springs Terrace. — How Formed. — As seen by Moonlight . . . 1

CHAPTER II.

Nature in Poetic Moods. — Is there Lurking Danger? — A Sanitarium. — The Liberty Cap. — The Giant's Thumb. — Singular Caves. — Falls of the Gardiner River. — In the Saddle. — Grand Cañon of the Yellowstone. — Far-Reaching Antiquity. — Obsidian Cliffs. — A Road of Glass. — Beaver Lake. — Animal Builders. — Aborigines of the Park. — The Sheep-Eaters. — The Shoshones and other Tribes 20

CHAPTER III.

Norris Geyser Basin. — Fire beneath the Surface. — A Guide's Ideas. — The Curious Paint Pot Basin. — Lower Geyser Basin. — Boiling Springs of Many Colors. — Mountain Lions at Play. — Midway Geyser Basin. — "Hell's Half Acre." — In the Midst of Wonderland. — "Old Faithful." — Other Active Geysers. — Erratic Nature of these Remarkable Fountains 34

CHAPTER IV.

The Great Yellowstone Lake. — Myriads of Birds. — Solitary Beauty of the Lake. — The Flora of the Park. — Devastating Fires. — Wild Animals. — Grand Volcanic Centre. — Mountain Climbing and Wonderful Views. — A Story of Discovery. — Government Exploration of the Reservation. — Governor Washburn's Expedition. — "For the Benefit of the People at Large Forever" 47

CHAPTER V.

Westward Journey resumed. — Queen City of the Mountains — Crossing the Rockies. — Butte City, the Great Mining Centre. — Montana. — The Red Men. — About the Aborigines. — The Cowboys of the West. — A Successful Hunter. — Emigrant Teams on the Prairies. — Immense Forests. — Puget Sound. — The Famous Stampede Tunnel. — Immigration 57

CHAPTER VI.

Mount Tacoma. — Terminus of the Northern Pacific Railroad. — Great Inland Sea. — City of Tacoma and its Marvelous Growth. — Coal Measures. — The Modoc Indians. — Embarking for Alaska. — The Rapidly Growing City of Seattle. — Tacoma with its Fifteen Glaciers. — Something about Port Townsend. — A Chance for Members of Alpine Clubs 73

CHAPTER VII.

Victoria, Vancouver's Island. — Esquimalt. — Chinamen. — Remarkable Flora. — Suburbs of the Town. — Native Tribes. — Cossacks of the Sea. — Manners and Customs. — The Early Discoverer. — Sailing in the Inland Sea. — Excursionists. — Mount St. Elias. — Mount Fairweather. — A Mount Olympus. — Seymour Narrows. — Night on the Waters. — A Touch of the Pacific 84

CHAPTER VIII.

Steamship Corona and her Passengers. — The New Eldorado. — The Greed for Gold. — Alaska the Synonym of Glacier Fields. — Vegetation of the Islands. — Aleutian Islands. — Attoo our most Westerly Possession. — Native Whalers. — Life on the Island of Attoo. — Unalaska. — Kodiak, former Capital of Russian America. — The Greek Church. — Whence the Natives originally came 109

CHAPTER IX.

Cook's Inlet. — Manufacture of Quass. — Native Piety. — Mummies. — The North Coast. — Geographical Position. — Shallowness of Behring Sea. — Alaskan Peninsula. — Size of Alaska. — A "Terra Incognita." — Reasons why Russia sold it to our Government. — The Price comparatively Nothing. — Rental of the Seal Islands. — Mr. Seward's Purchase turns out to be a Bonanza 127

CHAPTER X.

Territorial Acquisitions. — Population of Alaska. — Steady Commercial Growth. — Primeval Forests. — The Country teems with Animal Life. — A Mighty Reserve of Codfish. — Native Food. — Fur-Bearing Animals. — Islands of St. George and St. Paul. — Interesting Habits of the Fur-Seal. — The Breeding Season. — Their Natural Food. — Mammoth Size of the Bull Seals 143

CHAPTER XI.

Enormous Slaughter of Seals. — Manner of Killing. — Battles between the Bulls. — A Mythical Island. — The Seal as Food. — The Sea-Otter. — A Rare and Valuable Fur. — The Baby Sea-Otter. — Great Breeding-Place of Birds. — Banks of the Yukon River. — Fur-Bearing Land Animals. Aggregate Value of the Trade. — Character of the Native Race 159

CHAPTER XII.

Climate of Alaska. — Ample Grass for Domestic Cattle. — Winter and Summer Seasons. — The Japanese Current. — Temperature in the Interior. — The Eskimos. — Their Customs. — Their Homes. — These Arctic Regions once Tropical. — The Mississippi of Alaska. — Placer Mines. — The Natives. — Strong Inclination for Intoxicants . . . 173

CHAPTER XIII.

Sailing Northward. — Chinese Labor. — Unexplored Islands. — The Alexander Archipelago. — Rich Virgin Soil. — Fish Canning. — Myriads of Salmon. — Native Villages. — Reckless Habits. — Awkward Fashions and their Origin. — Tattooing Young Girls. — Peculiar Effect of Inland Passages. — Mountain Echoes. — Moonlight and Midnight on the Sea 186

CHAPTER XIV.

The Alaskan's Habit of Gambling. — Extraordinary Domestic Carvings. — Silver Bracelets. — Prevailing Superstitions. — Disposal of the Dead. — The Native "Potlatch." — Cannibalism. — Ambitions of Preferment. — Human Sacrifices. — The Tribes slowly decreasing in Numbers. — Influence of the Women. — Witchcraft. — Fetich Worship. — The Native Canoes. — Eskimo Skin Boats 199

CHAPTER XV.

Still sailing Northward. — Multitudes of Water-Fowls. — Native Graveyards. — Curious Totem-Poles. — Tribal and Family Emblems. — Division of the Tribes. — Whence the Race came. — A Clew to their Origin. — The Northern Eskimos. — A Remarkable Museum of Aleutian Antiquities. — Jade Mountain. — The Art of Carving. — Long Days. — Aborigines of the Yukon Valley. — Their Customs 212

CHAPTER XVI.

Fort Wrangel. — Plenty of Wild Game. — Natives do not care for Soldiers, but have a Wholesome Fear of Gunboats. — Mode of Trading. — Girls' School and Home. — A Deadly Tragedy. — Native Jewelry and Carving. — No Totem-Poles for Sale. — Missionary Enterprises. — Progress in Educating Natives. — Various Denominations engaged in the Missionary Work 222

CHAPTER XVII.

Schools in Alaska. — Natives Ambitious to learn. — Wild Flowers. — Native Grasses. — Boat Racing. — Avaricious Natives. — The Candle Fish. — Gold Mines Inland. — Chinese Gold-Diggers. — A Ledge of Garnets. — Belief in Omens. — More Schools required. — The Pestiferous Mosquito. — Mosquitoes and Bears. — Alaskan Fjords. — The Patterson Glacier 231

CHAPTER XVIII.

Norwegian Scenery. — Lonely Navigation. — The Marvels of Takou Inlet. — Hundreds of Icebergs. — Home of the Frost King. — More Gold Deposits. — Snowstorm among the Peaks. — Juneau the Metropolis of Alaska. — Ank and Takou Indians. — Manners and Customs. — Spartan Habits. — Disposal of Widows. — Duels. — Sacrificing Slaves. — Hideous Customs still prevail 246

CHAPTER XIX.

Aboriginal Dwellings. — Mastodons in Alaska. — Few Old People alive. — Abundance of Rain. — The Wonderful Treadwell Gold Mine. — Largest Quartz Crushing Mill in the World. — Inexhaustible Riches. — Other Gold Mines. — The Great Davidson Glacier. — Pyramid Harbor. — Native Frauds. — The Chilcats. — Mammoth Bear. — Salmon Canneries 258

CHAPTER XX.

Glacier Bay. — More Ice Bays. — Majestic Front of the Muir Glacier. — The Bombardment of the Glacier. — One of the Grandest Sights in the World. — A Moving River of Ice. — The Natives. — Abundance of Fish. — Native Cooking. — Wild Berries. — Hoouiah Tribe. — Copper Mines. — An Iron Mountain. — Coal Mines 275

CHAPTER XXI.

Sailing Southward. — Sitka, Capital of Alaska. — Transfer of the Territory from Russia to America. — Site of the City. — The Old Castle. — Russian Habits. — A Haunted Chamber. — Russian Elegance and Hospitality. — The Old Greek Church. — Rainfall at Sitka. — The Japanese Current. — Abundance of Food. — Plenty of Vegetables. — A Fine Harbor 293

CHAPTER XXII.

Contrast between American and Russian Sitka. — A Practical Missionary. — The Sitka Industrial School. — Gold Mines on the Island. — Environs of the Town. — Future Prosperity of the Country. — Hot Springs. — Native Religious Ideas. — A Natural Taste for Music. — A Native Brass Band. — Final View of the Capital 304

CHAPTER XXIII.

The Return Voyage. — Prince of Wales Island. — Peculiar Effects. — Island and Ocean Voyages contrasted. — Labyrinth of Verdant Islands. — Flora of the North. — Political Condition of Alaska. — Return to Victoria. — What Clothing to wear on the Journey North. — City of Vancouver. — Scenes in British Columbia. — Through the Mountain Ranges 321

CHAPTER XXIV.

In the Heart of the Rocky Mountains. — Struggle in a Thunder-Storm. — Grand Scenery. — Snow-Capped Mountains and Glaciers. — Banff Hot Springs. — The Canadian Park. — Eastern Gate of the Rockies. — Calgary. — Natural Gas. — Cree and Blackfeet Indians. — Regina. — Farming on a Big Scale. — Port Arthur. — North Side of Lake Superior. — A Midsummer Night's Dream 338

THE NEW ELDORADO.

CHAPTER I.

Itinerary. — St. Paul. — The Northern Pacific Railroad. — Progress. — Luxurious Traveling. — Riding on a Locomotive. — Night Experiences. — Prairie Scenes. — Immense Grain-Fields. — The Badlands. — Climbing the Rocky Mountains. — Cinnabar. — The Yellowstone Park. — An Accumulation of Wonders. — The Famous Hot Springs Terrace. — How Formed. — As Seen by Moonlight.

A JOURNEY from Massachusetts to Alaska was a serious undertaking a few years ago. It involved great personal risk, considerable expense, and many long months of weary travel; but it is now considered scarcely more than a holiday excursion, a good share of which may be denominated a marine picnic. That an important country, so easily accessible, should remain comparatively unexplored seems singular in the nineteenth century, especially when its great mineral wealth and natural attractions are freely admitted. The trip to Sitka, the capital of the Territory, and back is easily accomplished in three months, affording also ample time to visit the principal points of interest on the route, including the marvels of the Yellowstone National Park, in Wyoming, which is not only not surpassed in grandeur and beauty by any scenery on the continent, but in fact has no parallel on the globe. The traveler also naturally pauses on his way to examine at least one of the great mining centres of this gold-producing country, such as Butte, the "Silver City" of Montana, where he may behold scenes eclipsing in affluence the fabulous story of Midas. The plan adopted by the author, as herein detailed, was to make the westward journey by the Northern Pacific Railroad to Tacoma, on Puget Sound, where the remarkable inland sea voyage begins, thence sailing north to Pyramid Harbor and Glacier Bay, stopping as usual at the intermediate places of interest.

On the homeward passage, to vary the journey and to enjoy the wild scenery of British Columbia, Alberta, Assiniboia, and Manitoba, he left the steamer at Vancouver, returning by the Canadian Pacific Railway, which presents to the lover of nature such famous scenic advantages.

The journey westward seems practically to begin when the traveler reaches St. Paul, the capital of Minnesota, by way of Chicago, as here he strikes the trunk line of the Northern Pacific Railroad, which has an exclusive and unbroken track thence to Tacoma, a distance of nearly two thousand miles, the whole of which is covered with novelty and interest.

We will not pause to fully describe St. Paul, that youthful city of marvelous growth, promise, and beauty, with her mammoth business edifices of stone and brick, her palatial private residences, and her charming boulevards. The most casual visitor is eloquent upon these themes, as well as regarding the open-handed hospitality of her two hundred thousand inhabitants. Three iron bridges span the Mississippi at St. Paul, one of which is nearly three thousand feet long, supported upon arches two hundred and fifty feet in span, and having a roadway elevated two hundred feet above the water.

St. Paul is situated upon a series of terraces rising from the left bank of the Mississippi River, its site being both commanding and picturesque. Thus built at the head of navigation on a great waterway, it naturally commands a trade of no circumscribed character, besides enjoying the prestige of being the State capital.

Were it not for the unlimited facilities of transportation afforded by the grand and beneficent railroad enterprise embraced in the Northern Pacific system, the development of the vast and fertile country which lies between Lake Superior and the Pacific Ocean would have been delayed for half a century or more. It should be remembered that so late as 1850 there was not one mile of railroad in existence west of the Mississippi River. In 1836 there were, at most, but a thousand miles in operation on the entire American continent. This is an epoch of progress. Japan is traversed by railways, even China has caught the contagion, and is now building roads for the use of the iron horse in more than one direction within that ancient and widespread empire, while Russia and India are " gridironed " with rails.

It was remarked in a congressional speech in the year 1847 that the Rocky Mountains would be the limit of railroad enterprise across our continent; that the barrier presented by these huge elevations and the extensive "desert tract" beyond them must certainly prevent the development of the Pacific States.

"Desert," indeed!

No land on the globe produces such remarkable cereal crops as this very prairie soil is doing each successive year, not only supplying our own rapidly increasing population with the staff of life, but also feeding the less fortunate millions of Europe, where excessive labor and costly enrichment must make up the deficit arising from an exhausted soil and circumscribed area. The reader who follows these pages will not fail to see how liable legislators are to be mistaken in their predictions, and how apt events are to transcend the weak judgment of the confident and inexperienced declaimer. Even that Titan statesman, Daniel Webster, put himself on record in the United States Senate, while speaking against a proposition to establish a mail route through a portion of the western country, as follows: "What do we want with this vast, worthless area — this region of savages and wild beasts, of deserts of shifting sands and whirlwinds of dust, of cactus and prairie dogs? To what use could we ever hope to put these great deserts or those endless mountain ranges, impenetrable, and covered to their very base with eternal snow? What can we ever hope to do with the western coast, — a coast of three thousand miles, rock-bound, cheerless, uninviting, and not a harbor on it? What use have we for this country?"

In crossing the continent by the route we have chosen, one passes through a country whose grand scenic charms can hardly be exaggerated, in describing which superlatives only will apply, and whose agricultural advantages, natural resources, and mineral wealth are probably unequaled in the known world. We are taken through the productive wheat-fields of Minnesota and Dakota, among the gold and silver bearing hills of Idaho and Montana, into the prolific, garden-like valleys of Washington, whose lovely hopfields rival the gorgeous display of Kent in England, and whose abundant supply of coal and iron is only second to that of Pennsylvania.

The State has been, and may well be, denominated the Eden of the North Pacific.

On our way we are constantly meeting immense freight trains, laden with grain, flour, cattle, and other merchandise, bound for the Atlantic coast; long strings of coal cars, winding snake-like round sharp curves, and creeping up steep grades; passenger vans crowded with animated, intelligent people, all together testifying to the great and growing traffic of the West and Northwest. We pass scores of lofty grain elevators, high piles of lumber, and miles of various kinds of merchandise prepared for, and awaiting, shipment eastward, all of which evinces a local capacity for production far beyond our computation. How marvelous is the change from the conditions existing in this region a few years since, when millions of buffaloes roamed unmolested over these plains, valleys, and hills from Texas to Manitoba! The skeletons of these herds still sprinkle the prairies, bleached by the summer sun and crumbled by the winter's frost. Hundreds of carloads are annually shipped eastward to the factories which manufacture fertilizers.

As we speed on our western journey day and night, gliding through long tunnels and deep rock cuttings, over airy trestles, immense embankments, bridges, and viaducts, representing the skillful accomplishments of modern engineering, we carry along with us the domestic conveniences of home. The train, in fact, becomes our hotel for the time being, where we bathe, eat, sleep, and enjoy the passing scenery seated in luxuriously upholstered easy-chairs, which at night are ingeniously transformed as if by magic into soft and inviting beds. The elegance and comfort of these parlor, dining, and sleeping cars is calculated to make traveling what it has in a measure become, an inviting luxury. The miraculous cap of Fortunatus would seem to have been pressed into our service. So thoroughly perfected is the transcontinental railroad system that it is quite possible to enter the cars in an Atlantic city, say at Boston or New York, and not leave the train until five or six days have expired, when the objective point on the Pacific coast is reached.

While passing through deep gorges at night, or creeping over a mountain top, the effect from one's seat in the cars is weird and curious, especially when the winding track makes long curves in the train, so that the panting iron horse is seen from the rear, all ablaze and emitting dense clouds of smoke. The snow-tipped peaks on one side and the threatening gulch of unknown depth on the other assume a mantle of soft, gauze-like texture in the clear moonlight. At times one half believes the rails are laid upon the tree-tops, the branches of which loom up so close to us. Away in the valley, two thousand feet and more below our level, a rippling stream sparkles in the silvery light while on its way to swell some larger watercourse which drains the rocky hills. Looking far across the valley we try to make out the distant mountains, but only dim phantoms of gigantic size are seen, gliding stealthily away in the darkness.

We make interest with the conductor and engineer of the train for a special purpose. We are in search of a new sensation, to wit, such as may be derived from a night-ride on the engine, where one can see all the engineer sees, which is indeed little enough. The headlight of the locomotive throws its rays dimly on the darkness for a few rods in advance of the train. But what does that amount to, so far as being able to avoid danger? That brief space is passed in a second of time, and it is impossible to see what is beyond. The faithful engineer stands with both hands upon the machinery, one with which to instantly apply the brakes, the other to shut off the steam if danger shows itself ahead. That is all he can do. What a boisterous, asthmatic monster it is that drags the long train through the darkness at the rate of a mile in two minutes! How its hot breath belches forth, and how it springs and leaps over the iron track, fed incessantly with fresh fuel by the stoker! To one not accustomed to the oscillating motion, it is nearly impossible to keep his footing, much more difficult than on board of a pitching or rolling ship at sea. The motion is short, quick, and incessant. Black, — black as Erebus; how venturesome it seems to dash into such darkness! What a tempting of fate! Yet how few accidents, comparatively, occur! "The law of averages is what we calculate upon," said the engineer of No. —; "about so many people will be killed annually out of a given number of railroad travelers. We take all reasonable precautions to prevent accidents, but there are thousands of exigencies beyond our control." If any one proposes to you, gentle reader, to indulge in a night-ride on a locomotive, take our advice, and don't do it.

One does not linger in bed when passing through a country famous for its scenery. The experienced traveler has learned that the opening hours of the day are those in which his best and clearest impressions are received. He therefore rises betimes to enjoy the cool, dewy freshness of the morning. Now and again a prairie-owl is seen groping its winged way to shelter from the increasing light. He is sure to see plenty of coyotes, gray wolves, and graceful antelopes on the rolling prairies, each of these animals exhibiting in some special and interesting manner its natural proclivities. The prairie-dog nervously diving into and leaping out of its little prairie mound; the wolf bravely facing and glaring at the passing train, though careful to keep at a wholesome distance; and the antelopes in small herds hastening away by graceful bounds over the nearest hills, far too pretty and far too ornamental to shoot, suggesting in form and movements that most picturesque of wild animals, the Tyrolean chamois.

Minnesota presents to the eye of the traveler a grand and impressive country in the form of rolling prairies, diversified by lakes, — of which there are said to be ten thousand in the State, — forests, and inviting valleys, the latter particularly adapted for raising wheat and for dairy farming. Vast fields of ripening cereals are seen stretching for miles on either side of the railroad, without a fence to break their uniformity. This State possesses among other advantages that of a climate particularly dry, invigorating, and healthful. Four hundred miles of our route is through Northern Dakota, where the farming lands are easily tilled, well watered, and wonderfully prolific in crops. The choicest wheat grown in America, known as hard spring wheat, comes from this section, which has been called "the granary of the world." The gigantic scale on which wheat-raising is here conducted would seem incredible if faithfully described to an old-time New England farmer. The improvement which has been made in machinery connected with sowing, reaping, harvesting, and threshing grain enables one man to do as much in this western country as a dozen men could accomplish twenty-five or thirty years ago. There are wheat farms here embracing twenty thousand acres each, where economy in labor is of the utmost importance, and where the employees are so numerous as to be kept under semi-military organization. The author has seen the big grainfields of Russian Poland in their prime, but they are as nothing when compared with those of Northern Dakota, nor are the farming facilities which are generally employed throughout Europe nearly equal to those of this country.

At Bismarck, capital of the State, which is a small but energetic and thriving place, the Missouri River is crossed by a magnificent iron bridge, hung high in air, which cost a million dollars. This is the acme of successful engineering, passing our long, heavy train of cars over a track of gleaming rails from shore to shore without the least perceptible tremor, or the deflection of a single inch. The great waterway which it spans measures at this place fully twenty-eight hundred feet from bank to bank, though it is at this point two thousand miles from its confluence with the Mississippi.

The route we are following soon takes us through what are called the Badlands, a most singular region, where subterranean and surface fires are constantly burning, where trees have become petrified, and where the natural blue clay has been converted into terra cotta. This locality, extending for miles and miles, has been called Pyramid Park, on account of its fantastic forms presented in a singular variety of colors, and because of its mounds, domes, pyramids, and rocky towers. These vary as much in height as in form, some measuring ten feet, some two hundred, while all are clad in harlequin costume, black, white, blue, green, and yellow. It is called Badlands in contradistinction to the adjoining country, which is so very fertile, but the district is improved as good grazing ground for many thousands of cattle which supply our Atlantic cities with beef. Some of the best breeds of horses furnished to the Eastern States are raised, fed, and brought into marketable condition on these peculiar lands.

This region forms a sort of tangible hint of what we shall experience still farther on our Wonderland journey in the interesting and unequaled valley of the Yellowstone, where there are abundant evidences of volcanic force and subterranean fires, and where Nature is seen in her most erratic mood.

Just as we pass from Dakota into Montana, a short distance beyond the Little Missouri River, a lofty peak called Sentinel Butte is seen, at an elevation of nearly three thousand feet above sea level. The teeming, vigorous young life of the Northwest is manifest all along the route, with its wonderful energy and its almost incredible rate of progress. We were told that in the State which we had just left three thousand miles of railroad had been built and properly equipped before it contained a single town of more than five hundred inhabitants.

In the State of Montana we find a more hilly country than that through which we have so recently passed, yet it is well adapted to farming and possesses large areas of excellent grazing land. Indeed, there is scarcely any part of this territory, except the mountain ranges, where the climate is not sufficiently mild for cattle to winter out-of-doors. Undoubtedly they will thrive better for being housed at night in the coldest weather here or anywhere, but this is not absolutely necessary. No food is required for them except the native bunch grass, which cures itself, and stands as hay until the succeeding spring. Cattle are very fond of and will quickly fatten upon it. Sheep husbandry is also a great and growing interest here. We observe now and again a thrifty flock, tended by a boy-shepherd accompanied by his dog, recalling similar scenes in Tasmania and on the plains of Russia.

Statistics show that there are over two million acres now under cultivation in Montana, and that the territory is also fabulously rich in minerals. The present output of gold, silver, and copper is at the rate of three million dollars per month, and the yield of the mines is steadily on the increase.

As we hasten on our way, looking on one side

far down into sombre depths, and on the other at threatening, overhanging bowlders, or backward at the road-bed cut out of the solid rock which forms the cliff, we wonder at the successful audacity which conceived and built such a difficult highway. We have seen few instances of similar engineering so remarkable as is exhibited at certain points on the Northern Pacific Railroad. Equal difficulties have been overcome on the Zigzag Railway over the Blue Mountain Range, near Sidney, Australia, and also in Northern India, where the narrow gauge railroad climbs the foothills of the Himalayan Range to Darjeeling, about eight thousand feet above the plains of Hindostan, but in neither of these instances is the work so thorough, or on so gigantic a scale, as where the Northern Pacific crosses the Rocky Mountains.

We are quite conscious of being on an up grade, the large engine panting audibly from its extra exertion, and the train moving forward no faster than one could walk. Presently tall, snow-capped peaks come trooping into view, like mounted Bedouins clad in fleecy white, as the small city of Livingston is reached. This locality is about forty-five hundred feet above the sea. The town is situated in a beautiful valley, with nothing to indicate its altitude except the snow-crowned mountains not far away, standing like frigid sentinels. The observant traveler will also notice a certain rarefied condition of the atmosphere. Here we are about midway between the

Great Lakes and the Pacific coast, — between Superior, the largest lake on the globe, and the Pacific, the largest ocean in the world.

Livingston contains three thousand inhabitants, and is a thriving place, the frequent resort of many lovers of the rod and gun, both large and small game being found in abundance hereabouts. Forty miles north of Livingston is Castle Mountain mining district, rich in silver ores, and from whose argentiferous soil millions of dollars have been coined and hundreds of enterprising prospectors enriched. A branch road is taken at this point which runs directly southward to Cinnabar, a distance of nearly fifty miles, from which place coaches convey the traveler about six miles farther to the Wonderland of our continent, — the Yellowstone National Park.

The terminus of the railroad is known by the name of Cinnabar because it is situated at the base of a mountain bearing that title, remarkable for its exposure of vertical strata of three distinct geological periods. Here is a famous place known as the Devil's Slide, a singular formation caused by the washing out of a vertical stratum of soft material between one of quartzite and another of porphyry. The slide is two thousand feet high, and being of different color from the rest of the rocky mountain side is discernible for many miles away.

We have now reached one of the most remarkable points of our excursion, which demands more than a passing notice, sharing with the great glaciers of Alaska the principal interest of the present journey westward across the continent.

This magnificent territorial reservation is situated in the northwestern part of Wyoming, embracing also a narrow strip of southern Montana and southeastern Idaho, lying in the very heart of the Rocky Mountains. It was wisely withdrawn from settlement by an act of Congress in 1872, and is beneficently devoted forever to " the pleasure and enjoyment of the people." It forms a great preserve for wild animals, and a natural museum of marvels free to all. The well conceived liberality of this purpose is only commensurate with the unequaled grandeur of the Park itself, though at the time of passing this law comparatively little was actually known of the stupendous marvels contained within its widespread borders, besides which fresh discoveries of interest are still being made annually.

Of all those who have endeavored to depict this locality, none have been able to convey with the pen an adequate idea of its wild magnificence, or to give a satisfactory description of its acccumulated wonders. The eye alone can appreciate its indescribable beauty, majesty, and loveliness.

By the judicious expenditure of public money and the liberal outlay of corporate enterprise in road and bridge building, not to mention other facilities, one can now pretty thoroughly explore the Park in the brief period of a week or ten days. To do this satisfactorily heretofore required thrice this length of time, besides which, camping out was necessary; but it is no longer so, unless one chooses to play the gypsy. This plan is adopted by a few summer tourists, who take with them a regular camp outfit, depending upon the fish they catch for a considerable portion of their food supply during this out-of-door life.

The Park is under the control of the Secretary of the Interior. A local superintendent lives here, who is assisted by a few game-keepers and government police, besides which there is a small gang of laborers constantly at work during the favorable season, building roads and bridges, opening vistas here and there, and clearing convenient foot-paths, under the direction of an army engineer. Two companies of United States cavalry make their headquarters in the Park during the summer months, distributed so as to prevent any unlawful acts of visitors. The size of the reservation is sixty-four miles in length by fifty-four in width, thus giving it an area of over three thousand six hundred square miles. Or, to convey perhaps a clearer idea of its extent to the reader's mind, it may be said to be nearly one half the size of the State of Massachusetts. It is a volcanic region of incessant activity, with mountains ranging from eight to twelve thousand feet in height, and embracing a collection of spouting geysers, hot springs, steam holes, petrified forests, cascades, extraordinary cañons, and grand waterfalls, such as are unequaled in the known world.

We do not forget the well-known geysers of Iceland, or the Hot Lake district of New Zealand, with which the traveled visitor finds himself contrasting the phenomena of the Yellowstone.

The writer of these pages happened lately to see an article upon our National Park, written by the Earl of Dunraven, in which that gentleman questions whether the singular natural exhibitions here are not exceeded by those of New Zealand. We are familiar with both localities, and shall dismiss such a supposition simply by saying that the hot springs of the British colony referred to are no more to be compared with those of the Yellowstone Park, than is an artificial Swiss cascade comparable with Niagara. If Nature has anywhere else shown so wonderful a specimen of her handicraft, it has not yet been our lot to see it.

All the natural objects best worth visiting in the Park are now accessible by daily stages, which start at convenient hours from the hotel at Mammoth Hot Springs, making the round of the interesting sights; thus affording the general public every needed facility for examining the strangely attractive vicinity.

Near the hotel is an area of two hundred acres and more, covered here and there with boiling, terrace-building springs, which burst out of sloping ground in ceaseless pulsations, at an elevation of about a thousand feet above the Gardiner River near by, into which the main portion of the chemically impregnated waters flow. Five hundred feet from the base of the springs the water becomes cool, tasteless, and perfectly clear to the eye, as refreshing to drink as any water from the purest mountain rill. In ordinary quantities it has no evident medicinal effect, but is thought to be a wholesome tonic, with blood-purifying power. Some springs in the Park, though inviting in appearance, are to be avoided on account of certain objectionable medical properties which they possess. The hot springs adjacent to the hotel issue from many vents and at various elevations, slowly building for themselves terrace after terrace with circular pools, held in singularly beautiful stalactite basins, formed by depositing in thin layers the chemical substances which they contain. Some are infused with the oxide of iron, and produce a coating of delicately tinted red; others are exquisitely shaded in yellow by an infusion of sulphur; while some, from like causes, are of a dainty cream color. Upon numerous basins there are seen wavy, frill-like borders of bright green, indicating the presence of arsenic. Here and there the margins of the pools are scalloped and edged with a delicate bead-work, like Oriental pearls, while others are curiously honeycombed, and fretted with singular regularity. No artistic hand, however skillful, could equal Nature in these delicate and exquisitely developed forms. The grand terrace, viewed as a whole, is like a huge series of stairs or steps, two hundred feet high and five hundred broad, decked with variegated marble, together with white and pink coral. This immense calcareous formation might represent a frozen waterfall, or a congealed cascade. The water, in most instances, is at boiling heat as it pours out of the various openings, charged with iron, magnesia, sulphur, alumina, soda, and other substances. Every spring has its succession of limpid pools spreading out in all directions, the basins varying in size from ten to forty feet across their openings. When the sun penetrates the half enshrouding mist, and brings out the myriad colors of these beautiful terraces, the effect is truly charming; it is as though a rainbow had been shattered and the pieces strewn broadcast. While thus wreathed in vapors, as the evening approaches and the whole is touched by the rosy tints of the setting sun, the entire façade glows with softest opaline blushes, like a conscious maiden challenged by ardent admiration. For a moment, as we gaze upon its illumined expanse, it seems like a gorgeous marble ruin half consumed and still ablaze, the fire of which is being extinguished by an avalanche of snow-clouds. Such a scene cannot be depicted by photography; it cannot be represented faithfully by the artist's skillful touch in oils, because, like the vivid beauty of a sunset on the ocean, the light and shade are momentarily changing, while the prismatic hues gently dissolve into each other's embrace.

If possible, let the visitor witness the magic of the spot by moonlight. It is then fairy-like indeed, shrouded in a thin, silvery screen, — "mysterious veil of brightness made," — like the transparent yashmak of an East Indian houri.

CHAPTER II.

Nature in Poetic Moods. — Is there Lurking Danger? — A Sanitarium. — The Liberty Cap. — The Giant's Thumb. — Singular Caves. — Falls of the Gardiner River. — In the Saddle. — Grand Cañon of the Yellowstone. — Far-Reaching Antiquity. Obsidian Cliffs. — A Road of Glass. — Beaver Lake. — Animal Builders. — Aborigines of the Park. — The Sheep-Eaters. — The Shoshones and other Tribes.

How unapproachable is Nature in her poetic moods! how opulent in measure! how subtle in delicacy! No structure of truest proportions reared by man could equal the beauty of this lovely, parti-colored terrace. It recalled — being of kindred charm — that perfection of Mohammedan architecture the Taj-Mahal at Agra, as seen under the deep blue sky and blazing sun of India. Since the late sweeping destruction by earthquake and volcanic outburst of the similarly formed pink and white terraces in the Hot Lake district of New Zealand, at Tarawera, these of the Yellowstone Park have no longer a known rival. We may therefore congratulate ourselves in possessing a natural formation which is both grand and unique. In the far-away southern country referred to, there were no more symptoms foretelling the awful convulsion of nature which buried a broad, deep lake, together with an entire valley and native village, beneath lava and volcanic ashes, than there is exhibited in our own reservation at this writing. What signifies it that the Yellowstone Park has probably remained in its present comparatively quiet condition for many, many ages? The liability to a grand volcanic outburst at any moment is none the less imminent. History repeats itself. It has ever been the same with all great throes of Nature. Centuries of comparative quiet elapse, and then occurs, without any obvious predisposing cause, a great and awful explosion. The catastrophe of Pompeii is familiar to us all, which, in its turn, repeated the story of Herculaneum.

The Mammoth Hot Springs of the Yellowstone Park are not only beautiful in the tangible forms which they present, and the kaleidoscopic combinations of color which they produce, though their seeming crystal clearness is indescribable, but they have also remarkable medicinal virtues which enhance their interest and practical value. It is on this account that the place is gradually becoming a popular sanitarium, drawing patients from long distances at suitable seasons, especially those who suffer from rheumatic affections and skin diseases. Persistent bathing in the waters accomplishes many remarkable cures, if current statements can be credited, and there is ample reason for such a result. The pure air of this altitude must also be of great benefit to invalids generally, but more especially to those suffering from malarial poison and nervous prostration. The chemical properties of each spring are distinctive, most of them having been carefully analyzed, and the invalid is thus enabled to choose the one which is presumably best adapted to his special ailment.

Groups of pines, or single trees, find sufficient nutriment in the calcareous deposit to support life, and thus a certain barrenness is robbed of its depressing effect, while the whole is partially framed by densely wooded hills which serve to throw the terraces strongly into the foreground. When we last looked upon the scene the sun was setting amid a canopy of gold and orange hues, as the evening gun of the military encampment in the valley echoed again and again in sonorous tones among the everlasting hills, and died away in the distant gorges of the Yellowstone.

A lady visitor who entered the Park at the same time with the author, on the first day of her arrival placed a pine cone in one of the springs near to the hotel. So rapid is the action of the mineral deposit which is constantly going on that at the close of the eighth day the cone was taken from the spring crystallized, as it were, being encrusted with a silicious deposit nearly the sixteenth of an inch in thickness. Branches of fern, acorns, and other objects are treated in a similar manner, often producing very charming and peculiar ornaments which serve as pleasing souvenirs of the traveler's visit.

In sight of the hotel piazza there is a curious and interesting object, built up by a spouting spring long since extinct, and which has been named the Liberty Cap. It is a little on one side but yet in front of the terraces, and appears to be composed entirely of carbonate of lime. With a diameter of about fifteen feet at the base, it gradually tapers to its apex forty feet from the ground. This prominent formation, though remarkable, is yet no mystery. It was produced by the waters of a spring, probably forced up by hydrostatic pressure, overflowing and precipitating its sediment around the vent, until finally, the cause ceasing, the pressure become exhausted and the cone was thus formed. It may have required ages of activity in the spring thus to erect its own mausoleum, — no one can safely conjecture how long. Still nearer to the terraces is a similar formation called the Giant's Thumb. Both are slowly becoming disintegrated by atmospheric influences; we say slowly, since they may still exist, slightly diminished in size, a hundred years hence. There is manifestly a tendency in the springs which are now active in other parts of the neighborhood to build just such tall cylinders of sinter about their vents. Some of the partially formed cones in the vicinity are perfect, as far as they have accumulated, while others present a broken appearance, as if shattered by a sudden explosion.

There are several caves in the neighborhood of the terraces daintily ornamented with stalactites of snowy whiteness, where springs which have long since become exhausted were once as active as those which now render this place so interesting. From one of these caves there issues a peculiar gas, believed to be fatal to animal life. A bird, it is said, flying across the entrance close enough to inhale the vapor will drop lifeless to the ground. We are not prepared to vouch for this, — indeed we very much doubt the guide's story, — but it naturally recalled the Grotto del Cane, near Naples, where it will be remembered the guides are only too ready to sacrifice a dog for such visitors as are cruel enough to permit it, by causing the animal to inhale the poisonous gas which settles to the lower part of the cave so named.

There is another cave not far from the hotel very seldom resorted to, and which appears to have once been the operating sphere of a large geyser, but which is now only a dark hole. Into this one descends by a ladder. It is a weird, uncanny place, requiring torches in order to see after entering its precincts. Aroused by the artificial light, myriads of bats drop from the ceiling, until the place seems alive with them. Now and then in their gyrations one touches the visitor's hand or cheek with its cold, damp body, causing an involuntary shudder. Verily, the Bats' Cave is not an inviting place to visit.

One of the first places which the stranger seeks after enjoying the attractions of the terraces and a few curiosities near to the hotel is the Middle Falls of the Gardiner River, situated three or four miles away in a southerly direction. Here we look down into a broad, dark cañon considerably over a thousand feet deep, and whose rough, precipitous sides are nearly five hundred feet apart at the summit, gradually narrowing towards the bottom. The Gardiner River flows through the gorge, having at one place an unbroken fall of a hundred feet; also presenting a mad, roaring, rushing series of cascades of three hundred feet descent. The aspect and general characteristics of this turmoil of waters recalled the famous Falls of Trolhätta, in Sweden. The hoarse music of the waters, rising through the branches of the pines which line the gorge, pierce the ear with a thrilling cadence all their own, while the dark cañon stretches away for many miles in its wild and sombre grandeur. It is well to visit this spot before going to greater distances from the hotel. Impressive as it is sure to prove, there is yet a much superior feature of the Park, of similar character, which remains to be seen. We refer to the Grand Cañon of the Yellowstone River, where an immense cataract is formed by the surging waters near the head of the gorge, which here narrows to about one hundred feet. The volume of water is very great at the point where it rushes over a ledge nearly four hundred feet in height, at one bold leap. This is known as the Lower Fall, there being another half a mile above it, called the Upper Fall, which is one hundred and fifty feet high. These falls are more picturesque, but less grand than the Lower. They are presented to our view higher up among the green trees, where lovely wild flowers and waving ferns cling to the rocks, and under the inspiring rays of the sunlight add to their brightness and crystal beauty. A waterfall, like an oil-painting, may be hung in a good or a disadvantageous position as to light, and both are largely dependent upon this contingency for their inspiring charm.

The Great or Lower Fall of the Yellowstone Cañon is twice as high as Niagara, while the beautiful blazonry on the walls of the deep gorge, like some huge mosaic, all aglow with matchless color, marvelous in opulence, adds a fascinating charm unknown to the mammoth fall just named. These varied hues have been produced by the snow and frost, vapor and sunshine, the lightning and the rain of ages, acting upon certain chemical constituents of the native rock. This is said to be the most wonderful mountain gorge, when all of its belongings are taken into consideration, yet discovered. It is over twenty miles long, and is in many places from twelve to fifteen hundred feet deep. The author has visited the imposing cañons of Colorado, the thrilling gorges of the Yosemite, and some of still greater magnitude in the Himalayan range of northern India, but never has he seen the equal of this Grand Cañon of the Yellowstone, or beheld so high a waterfall of equal volume.

A safe platform has been erected at the edge of the fall, where one can stand and witness its amazing plunge of over three hundred and fifty feet. The stranger instinctively holds his breath while watching the irresistible volume of water as it advances, and follows it with the eye into the profound depth of the cañon. The best view of the gorge, however, is that obtained from Lookout Point, situated about a mile south of the Lower Fall. A half mile farther in the same direction, and at the same elevation, lies Inspiration Point, from whence a more comprehensive outlook may be enjoyed. The grouping of crags, pinnacles, and inaccessible points is grand and inexpressibly beautiful. Eagles' nests with their young are visible at eyries quite out of reach, save to the monarch bird itself. On other isolated points, far below us, are seen the nests of fish-hawks, whose builders look like swallows in size as they float upon the air, or dart for their prey into the swift, tumultuous stream that threads the valley. Gazing upon the scene, the vastness of which is bewildering, a sense of reverence creeps over us, — reverence for that Almighty hand whose power is here recorded in such unequaled splendor. At last it is a relief to turn away from looking into the sheer depth and reach a securer basis for the feet. Still we linger until the sunset shadows lengthen and pass away, followed by the silvery moonlight. Every hour of the day has its peculiar charm of light and shade as seen upon the cañon and its churning waters.

The excursion out and back from the hotel to view the principal points of interest in the neighborhood covers a distance of about seven miles through the woods and along the threatening brink of the gorge. A rude Indian trail affords the only means of reaching the several outlooks. Saddle-horses are supplied for the excursion by the hotel proprietor, and visitors generally avail themselves of this mode of transportation. The horses employed for the service are remarkably sagacious and sure-footed. Understanding exactly what is required of them, they overcome the deep pitches and abrupt rises of the narrow, tortuous way with great ingenuity and caution. At times one is borne so near the brink of the awful chasm as to make the passage rather exciting. It must be admitted that a single misstep on the part of the animal which bears him would hurl horse and rider two thousand feet down the cañon to instant destruction. There is no barrier between the cliff and the few inches of earth forming the path. Visitors are cautioned at starting to give the horses their heads, and not attempt to guide them as they would do under ordinary circumstances. The intelligent animals fully comprehend the exigencies of the situation. On the occasion of the writer's visit the equestrian party consisted of nine persons, including the guide; of these, two ladies and one gentleman abandoned the saddles after the first mile, finding the seeming danger too much for their nerves, and completed the long tramp on foot.

"What wonderful majesty and beauty are hidden here from an unconscious world," said an experienced member of our little party whom chance had brought together at the brink of the gorge. "Everybody visits Niagara," he continued, "but few, comparatively, participate in the glory and loveliness of this place, and yet how superior in attraction it is to those lines of summer travel, the Natural Bridge of Virginia, the Mammoth Cave of Kentucky, or even the justly famed Yosemite Valley;"— a sentiment which all heartily indorsed.

In these pages we pass rapidly from one great attraction to another, because we have only a limited space in which to speak of them, but the intelligent and appreciative visitor will be more leisurely in his examination. Hours may be profitably occupied in the careful observation and thorough enjoyment of each locality, the interest growing by what it feeds upon. One hardly realizes the passage of time when occupied in the contemplation of such strange and absorbing objects, and is apt to linger thoughtfully until he is warned by the business-like suggestion of the guide.

Another interesting spot which the stranger will hasten to visit is the Obsidian Cliffs, situated about a dozen miles from the hotel. These singular and, so far as we know, unique cliffs are formed of volcanic glass, and measure a thousand feet in length by nearly two hundred in height, recalling in general effect the Giant's Causeway in the north of Ireland. They rise in almost vertical columns from the eastern shore of Beaver Lake. The color of the glass is dark green, like that of which cheap quart bottles are made, and though the glass glistens like jet it is opaque. A carriage road has been provided, — a glass road, — a quarter of a mile long, running by the base of the cliffs. To construct this road large fires were built upon the obsidian mass, which, when thoroughly heated, was dashed with cold water, causing it to crack and crumble to pieces. It was a tedious undertaking, but an available roadway was at last the result.

Close at hand is Beaver Lake, of artificial origin, having been created by the industrious animal after which it is named. A colony have here built a series of thirty dams, thus forming a sheet of water of considerable depth, half a mile in width, and two miles long, framed by tall, straight pines, and covered near the shore with aquatic flowers. As we passed the lake, in its shady corners were seen flocks of ducks in gaudy colors and of many different species, while on the far side representatives of the beaver tribe were kind enough to exhibit themselves for our amusement. The series of dams which these little creatures have constructed hereabouts have falls of from three to six feet each, extending for a distance of nearly two miles. The lily plants which bordered Beaver Lake were of a curious amber color, growing here and there in groups of great density. At a snap of the driver's whip a bevy of wild ducks rose, but lazily settled again upon the water close at hand. "They have read the printed regulations of the Park," said the driver, "and know that no one will attempt to shoot them." Beyond the lake are broad patches of level meads, sprinkled with lovely wild flowers, in which yellow, purple, and white prevailed. The delicate little phlox, modestly clinging to the ground, was fragrant above all the rest. Occasional spots bordering the pine woods showed the exquisite enamel of the blue violets, which emitted their familiar and welcome fragrance. These were dominated by a tall, regal flower, clustering on one stem, whose name we know not, but which formed great masses of purple bloom.

Close to the curious and interesting Obsidian Cliffs is a pleasant resort called Willow Park, a cool, shady spot, where a clear stream of good water flows through a stretch of rich pasture land, forming a delightful rural picture, full of peaceful and poetic suggestiveness. This is a favorite camping ground for those who adopt that mode of visiting the several sections of the Park.

The stranger looks about him in silent amazement, wondering how long Nature has been displaying her erratic moods after the fashion exhibited here, now smiling with winning tenderness, and now frowning with implacable sternness. He sees everywhere evidences of great antiquity, and beholds objects which must date from time incalculably remote, but there is no recorded history extant of this strange region. The original Indian inhabitants of the Park were a very peculiar people, — a sort of gnome race, — a tribe individually of Liliputian size, who lived in natural caves, of which there are many in the hills, where rude and primitive implements of domestic use belonging to the aborigines have been found. They do not seem to have possessed even the customary legends of savage races concerning their surroundings and their origin. This tribe, the former dwellers here, were called the Sheep-eating Indians, because they lived almost solely upon the flesh, and clothed themselves in the skins, of the big-horn sheep of these mountains, — an animal which is found running wild in more or less abundance throughout the whole northern range of the Rocky Mountains, even where it reaches into Alaska. These natives are represented to have been a timid and harmless people, without iron tools or weapons of any sort, except bows and arrows, to which may be added hatchets and knives formed of the flint-like volcanic glass indigenous to the Park. They were an isolated people from the very nature of their country, which was nearly inaccessible at all seasons, and entirely so during the long and severe winters.

Other native tribes were debarred from this region through superstitious fear, induced by the incomprehensible demonstrations of Nature exhibited in boiling springs, spouting geysers, and the trembling earth, accompanied by subterranean explosions. This seemed to them to be evidence of the wrath of the Great Spirit, angered, perhaps, by their unwelcome presence. The Sheep-eaters, born among these scenes, gave no special heed to them, and rather fostered an idea which prevented others from interfering with the surrounding game, and which also gave them immunity from the otherwise inevitable oppression of a stronger and more aggressive people than themselves. As civilization advanced westward, or rather as the white man found his way thither, this Yellowstone tribe gradually dwindled away or became united with the Shoshones of Iowa. Their individuality seems now to have been entirely lost, not a trace of them, even, being discernible, according to more than one intelligent writer upon the subject.

No Indians of any tribe are now permitted in the reservation, otherwise, lazy as these aborigines are, they would soon make reckless havoc among the fine collection of wild animals which is gathered here. The Indians are all in the annual receipt of money and ample food supplies from the government; and the killing of extra game and selling the hides would furnish them with only so many more dollars to be expended for whiskey and tobacco. These tribes have no idea of economy, or care for the future. The reliance they place upon government supplies promotes a spirit of recklessness and extravagance. If their potato crop fails, or partial famine sets in from some extraordinary cause, it finds them utterly unprepared to meet the exigency. Oftentimes it is found that the government rations and supplies have been sold, and the money received therefor lavishly squandered.

CHAPTER III.

Norris Geyser Basin. — Fire beneath the Surface. — A Guide's Ideas. — The Curious Paint Pot Basin. — Lower Geyser Basin. — Boiling Springs of Many Colors. — Mountain Lions at Play. — Midway Geyser Basin. — " Hell's Half Acre." — In the Midst of Wonderland. — " Old Faithful."— Other Active Geysers. — Erratic Nature of these Remarkable Fountains.

A PLEASANT drive of twenty miles in a southerly direction from the Hot Springs Hotel, through the wildest sort of scenery, over mountain roads and beside gorgeous cañons, will take the visitor to the Norris Geyser Basin, a spot which promptly recalled to the writer somewhat similar scenes witnessed at the aboriginal town of Ohinemutu, in the northern part of New Zealand. Clouds of sulphurous vapor constantly hang alike over both places, produced by a similar cause, though the scene here is far more vivid and demonstrative. This whole basin is dotted by hot water springs and fumaroles, which maintain an incessant hissing, spluttering, and bubbling, night and day, through the twelve months of the year. The water which issues from these sources is of various colors, according to the impregnating principle which prevails, the yellow sulphur vats being especially conspicuous to the sight and offensive to the smell. What a strange, weird place it is! No art could successfully imitate these extravagances of Nature. Some of the rills are cool, others are boiling hot; some are white, some pink or red, and one large basin, fifty feet across, is called the Emerald Pool, because of its intensely green color; yet it appears to be quite pure and transparent when a sample is taken out and examined. Each spring seems to be entirely independent of the rest, though all are situated so near to each other. An almost constant tremor of the earth is realized throughout this immediate region, as though only a thin crust separated the visitor from an active volcano beneath his feet; and, notwithstanding the various scientific theories, who can say that such is not actually the case?

"I know all about the idea that these eruptions of boiling water, steam, and sulphurous gases are produced by chemical action," said our guide. "I've heard lots of scientific men talk about the subject, but I don't believe nothing of the sort."

"And why not?" we asked.

"Do you believe," he said, "that chemical action in the earth could create power enough, first to bring water to 212° of heat, and then force it two hundred feet into the air a number of times every day in a column four or five feet in diameter, and keep it up for quarter of an hour at a time?"

"Well, it does seem somewhat problematical," we were forced to answer.

"After living here summer and winter for six years," he said, "I have seen enough to satisfy me that there is a great sulphurous fire far down in the earth below us, which, if the steam and power it acccumulates did not find vent through the hundreds of surface outlets distributed all over the Park, would seek one by a grand volcanic outburst."

"Put your hand on the ground just here," he continued, as we walked over a certain spot where our footfall caused a reverberation and trembling of the soil.

"It is almost too hot for the flesh to bear," we said, quickly withdrawing our hand.

"Too hot! I should say so. Now I don't believe anything but a burning fire can produce such heat as that," he added, with an expression of the face which seemed to imply, "I don't believe you do either."

"The original volcanic condition of this whole region seems also to argue in favor of your deductions," we replied.

"That's just what I tell 'em," continued the guide. "Them big fires that first did the business for this neighborhood are still smouldering down below. You may bet your life on that."

This rather startling idea is emphasized by a smoking vent close at hand, which is also constantly sending forth superheated steam and sulphurous gases, like the extinct volcano of Solfatara, near Naples. Sulphur crystals strew the ground, and are heaped up in small yellow mounds. Not far away an intermittent geyser bursts forth every sixty seconds from a deep hole in the rock-bed of the basin, showing a stream of water six inches in diameter, and sending the same skyward thirty or forty feet. Here also is a powerful geyser called the Monarch, which leaps into action with great regularity once in twenty-four hours, throwing a triple stream to the height of a hundred and thirty feet, and continuing to do so for the space of fifteen or twenty minutes. Beneath the sun's rays all the colors of the prism are reflected in this vertical column of water, and not infrequently the distinct arch of a rainbow is suspended like a halo about its crown. Nature, even in her most fantastic caprice, is always beautiful.

There are several other high-reaching and powerful geysers in this vicinity, but we will not weary the reader by pausing to describe them.

Gibbon Paint Pot Basin is next visited, being a most curious area, measuring some twenty acres, more or less, situated in a heavily-wooded district, not far from Gibbon Cañon. Here is a most strange collection of over five hundred springs of boiling, splashing, exploding mud, exhibiting many distinct colors, which gives rise to the name it bears. One pot is of an emerald green, another is as blue as turquoise, a third is as red as blood, a fourth is of orange yellow, another is of a rich cream color and consistency. The visitor is struck by the singularity of this hot-spring system, which produces from vents so close together colors diametrically opposite. The earth is piled up about the seething pools, making small mounds all over the basin, and forming a series of pots of clay and silicious compounds. Near the entrance of Gibbon

Cañon is a remarkable collection of extinct geysers; the tall, slim, crystallized structures, originating like the Liberty Cap already described, look like genii totem poles, corrugated by the finger of time, and forming significant monuments of bygone eruptions, while the surrounding volcanoes were slowly exhausting their fury. Even about these long-extinct geysers there is an atmosphere indicating their former intensity, though it is quite possible they may have been sleeping for ten centuries.

The locality known as the Lower Geyser Basin is filled with striking and somewhat similar volcanic exhibitions, though there are more hot springs here than other phenomena, the aggregate number being a trifle less than seven hundred, including seventeen active geysers. In some respects this spot exceeds in interest those previously visited, being more readily surveyed as a whole. The variety of form and the large number of these springs are remarkable. As a rule they are less sulphurous and more silicious than those already spoken of. Here, as at the terraces near the hotel, the last touch of beauty is imparted by the sun's rays forcing themselves through the white vapory clouds which are thrown off by the mysteriously heated waters. One of the large basins, measuring forty by sixty feet, is filled with a sort of porcelain slime, notable for its soft rose tints and delicate yellow hues, which are brought out with magic effect under a cloudless sky. This basin has an elevation of over seven thousand feet above the level of the sea, and is surrounded by heavily-timbered hills which are four and five hundred feet higher. Numerous as these springs and geysers are, each one is strongly individualized by some special feature which marks it as distinctive from the rest, and renders it recognizable by the residents of the Park, but which, however interesting to the observing visitor, would only prove to be tedious if here described in detail.

While sitting at twilight on the piazza of the rude little inn where we passed the night in this basin, there came out from the edge of the wood on to a broad green plateau a couple of long tailed mountain lions. They were not quite full grown, and were of a tawny color. These creatures, savage and dangerous enough under some circumstances, seemed half tame and entirely fearless, playfully romping with each other, and exhibiting catlike agility. The proprietor of the inn told us that not long since, upon a dark night, they came to the house and attacked his favorite dog, killing and eating him, leaving only the bones to explain his disappearance in the morning. They, too, must have read the regulations, "No firearms permitted in the Park."

The Midway Geyser Basin is situated a few miles directly south of that just spoken of, and contains an extraordinary group of hot springs, among which is the marvelous Excelsior Geyser, largest in the known world. It bursts forth from a pit two hundred and fifty feet in diameter, worn in the solid rock, and which is at all times nearly full of boiling water, above which there is constantly floating a dense column of steam, which rising slowly is borne away and absorbed by the atmosphere. The water which flows so continuously over the brim has formed a series of terraces beaming with beautiful tints. This stupendous fountain is intermittent, giving an exhibition of its startling powers at very irregular periods, when it is said to send up a column of water sixty feet in diameter to a height of from fifty to one hundred feet! So great is the sudden flood thus produced in the Firehole River, which is here between seventy-five and a hundred yards broad, that it is turned for the time being into a furious torrent of steaming, half-boiling water. The Excelsior has also a disagreeable and dangerous habit of throwing up hundred-pound stones and metallic débris with this great volume of water, while the surrounding earth vibrates in sympathy with the hidden power which operates so mysteriously. Visitors naturally hasten to a safe distance during these moments of extraordinary activity.

About midway between Firehole and the Upper Geyser Basin is a strange, unearthly, vaporous piece of low land, which is endowed with a name more expressive than elegant, being called "Hell's Half Acre." Here again it seems as if this spot is separated from the raging fires below by only the thinnest crust of earth, through which numerous boiling springs find riotous vent. The soil in many parts is burning hot, and echoes to the tread as though liable to open at any moment and swallow the venturesome stranger. During the season of 1888, a lady visitor who stepped upon a thin place sank nearly out of sight, and though instantly rescued by her friends, she was so severely scalded as to be confined to her bed for a month and more at the Mammoth Springs Hotel. The air is filled with fumes of sulphur, and the place would seem to be appropriately named. There are forty springs in this " Half Acre," which, by the way, occupies ten times the space which the name indicates, where the seething and bubbling noise is like the agonized wailing of lost spirits. The place has another, and perhaps better, designation besides this satanic title, namely, Egeria Springs. Great is the contrast between the heavens above and the direful suggestions of the earth below, as we behold it under the serene beauty of the blue sky which prevails here in the summer months, and which renders camping out in the Park delightful. " You should come here during a thunder-storm," said our companion, who is a dweller in this region. " I have done so twice," he continued, " simply to witness the fitness of the association : rolling thunder overhead and flashes of lightning in the atmosphere, through which the boiling vats, hissing pools, and steaming fissures are seen in full operation, as though they were a part and parcel of the electric turmoil agitating the sky."

It is impossible to appreciate these various phenomena in a single hurried visit. Like the Falls of Niagara, or the Pyramids of Gizeh, they must become in some degree familiar to the observer before he will be able to form a complete, intelligent, and satisfactory impression which will remain with him. One cannot grasp the full significance of such accumulated wonders at sight. We look about us among the green trees that border the open areas, surprised to behold the calm sunshine, the tuneful birds, and the chattering squirrels, moved by their normal instincts, utterly regardless of these myriad surrounding marvels.

The grandest spouting springs are to be found in Upper Geyser Basin, where there are twenty-five active fountains of this character. Here is situated the famous "Old Faithful," which, from a mound rising gradually about six or eight feet above the surrounding level, emits a huge column of boiling water for five or six minutes in each hour with never-failing regularity, while it gives forth at all times clouds of steam and heated air. The height reached by the waters of this thermal fountain varies from eighty to one hundred and twenty feet, and it has earned its expressive name by never failing to be on time. It seemed, somehow, to be a more satisfactory representative of the spouting spring phenomenon than any other in the entire Park, though it would be difficult to say exactly why. Its prominent position, dominating the rest of the geysers of the basin, gives it special effect. Irrespective of all other similar exhibitions, the stately column of "Old Faithful" rises heavenward with splendid effect in the broad light of day, or in the still hours of the night, once in every sixty minutes, as uniformly as the rotation of the second-hand of a watch. The effect was ghostly at midnight under the sheen of the moon and the contrasting shadows of the woods near at hand, while not far away, across the Firehole River, the lesser geysers were exhibiting their erratic performances, casting up occasional crystal columns, which glistened in the silvery light like pendulous glass. There is quite a large group of geysers in this immediate vicinity, which perform with notable regularity at stated periods. There is one called the Beehive, because of its vent, which has a resemblance to an old-fashioned straw article of the sort, the crater being about three feet in height. The author saw this spring throw up a stream three feet in diameter nearly or quite two hundred vertical feet for eight or ten minutes, when it gradually subsided. There are over four hundred geysers and boiling springs in this basin. Among them is the Giantess, situated four hundred feet from the Beehive, which does not display its powers oftener than once in ten or twelve days; but when the eruption does take place, it is said to exceed all the rest in the height which it attains and the length of time during which it operates. It has no raised crater, but comes forth from a vent even with the surface of the ground, thirty-four feet in length and twenty-four in width. When it is in action, so great is the force expended that miniature earthquakes are felt throughout the immediate neighborhood. There are seen, not far away, the Lion, Lioness, Young

Faithful, the Grotto, the Splendid, etc., each one more or less operative. We have by no means enumerated all the active fountains in this basin, seeking only to designate their general character. However well prepared for the outburst, one cannot but feel startled when a geyser suddenly rises, mysteriously and ghost-like, close at hand, from out the deep bowels of the earth, its white form growing taller and taller, while the spray expands like weird and shrouded arms. To heighten this sepulchral effect the atmosphere is full of sulphurous vapors, while strange noises fall upon the ear like subterranean thunder. What puzzling mysteries Nature holds concealed in her dark, earthy bosom!

Let us not forget to mention, in this connection, one of the largest fountains of the Firehole Basin, namely, the Grand Geyser, which is placed next to the Excelsior in size and performance. This fountain has no raised cone, and operates once in about thirty-six hours. Of course the visitor is not able to see each and all of these strange fountains in operation. He might remain a month upon the ground and not do so; consequently, he is obliged to take some of the dimensions and performances on trust; but most of the statements which are made to him can easily be verified.

When this Grand Geyser is about to burst forth, the deep basin, which is twenty feet and more across, first gradually fills with furiously boiling water until it overflows the brim;' then it becomes shrouded by heavy volumes of steam, out of which come several loud reports, like the discharge of a small cannon, when suddenly the whole body of water is lifted, and a column ten or twelve feet in diameter rises to a height of ninety feet, from the apex of which a lesser stream mounts many feet higher, until the earth trembles with the force of the discharge and falling water as it rushes towards the river. This strange exhibition lasts for eight or ten minutes, then the fountain slowly subsides, with hoarse mutterings, like some retreating and overmastered wild beast, growling sullenly as it disappears.

It will thus be seen that these geysers vary greatly in their action, in the duration of their eruptions, and in the intervals which elapse between the performances. Some of them labor as though the water was slowly pumped up from vast depths, some burst forth with full vigor to their highest point at once, while others become exhausted with a brief effort. There are a few that subside only to again commence spouting, being thus virtually continuous; but these are not of such power as to throw their streams to a great height. One group of this sort is called the Minute Men, some of which spout sixty times within the hour; others eject small streams incessantly.

This immediate valley is very irregular in surface and thickly wooded in parts, showing also the ruins of many extinct geysers. It is a dozen miles long and between two and three wide, literally crowded with wonders from end to end. It contains a collection of boiling and spouting springs on a scale which would belittle all similar phenomena of the rest of the known world, could they be brought together.

As the reader will have understood, the period of activity with all the geysers is more or less irregular, except in the instance of Old Faithful. We have no knowledge of a simultaneous eruption having ever taken place. Many of these active springs which now exist will, doubtless, sooner or later subside and new ones will form to take their places, a process which has been going on, no one can even guess for how many ages.

CHAPTER IV.

The Great Yellowstone Lake. — Myriads of Birds. — Solitary Beauty of the Lake. — The Flora of the Park. — Devastating Fires. — Wild Animals. — Grand Volcanic Centre. — Mountain Climbing and Wonderful Views. — A Story of Discovery. — Government Exploration of the Reservation. — Governor Washburn's Expedition. — "For the Benefit of the People at Large Forever."

In the southern section of the Yellowstone Park, near its longitudinal centre, is one of the most beautiful yet lonely lakes imaginable, framed in a margin of sparkling sands, and surrounded by Alpine heights. One stretch of the shore about five miles long is called Diamond Beach; the volcanic material of which it is formed, being entirely obsidian, reflects the sun's rays like brilliant gems, while the beach is caressed by wavelets scarcely less bright. Surrounded by many wonders, the lake is itself a great surprise, lying in the bosom of rock-ribbed mountains at an elevation of nearly eight thousand feet above the sea. We know of but one other large body of water on the globe at any such height, namely, Lake Titicaca, in South America, famous in Peruvian history. The Yellowstone Lake is always of crystal clearness, and is fed from the eternal snow that piles itself up on the lofty peaks which surround it, and which are sharply outlined in all directions against the blue of the sky. The outlet of the lake is the Yellowstone River, which issues from the northern end, while the Upper Yellowstone runs into it on the opposite side. The lake is twenty-two miles long by fifteen in width, and has an area of a hundred and fifty square miles. Its greatest depth is three hundred feet, and it is overstocked with trout, many of which, unfortunately, are infested by a parasitic worm which renders them unfit for food; but this is not the case with all the fish; a large portion are good and wholesome. Geologists find sufficient evidence to satisfy them that this lake, now narrowed to the dimensions just given, in ancient times covered two thirds of the present Park. Aquatic birds abound upon its broad surface, and build their myriad nests on its green islands. They are of many species, comprising geese, cranes, swans, snipe, mallards, teal, curlew, plover, and ducks of various sorts. Pelicans swim about in long white lines; herons, in their delicate ash-colored plumage, stand idly on the shore, while ermine-feathered gulls fill the air with their loud and tuneless serenade. Hawks, kingfishers, and ravens also abound on the shore, the first-named watching other birds as they rise from the water with fish, which they make it their business, freebooter-like, to rob them of. The lake has many thickly-wooded islands, and there are several long, pine-covered promontories which stretch out in a graceful manner from the mainland, the whole forming a grand primeval solitude. Now and again a solitary eagle, on broad-spread pinions, sails away from the top of some lofty pine on the mountain side to the deep green seclusion of the nearest island. Even the presence of this proud and austere bird only serves to emphasize the grave and solemn loneliness which rests upon the locality.

It is a charming feature of this placid lake which causes it to gather into its bosom a picture of all things far and near: the clouds, "those playful fancies of the mighty sky," seem to float upon its surface; the blue of the heavens is reflected there; the tall peaks and wooded slopes mirror themselves in its depths. As we look upon the lake through the purple haze of sunset, a picture is presented of surpassing loveliness, tinted with blue and golden hues, which creep lovingly closer and closer about the quiet isles; while there come from out the forest resinous pine odors, delightfully soothing to the senses, accompanied by the soft music of swaying branches, and the low drone of insect life.

To linger over such a scene is a joy and an inspiration to the experienced traveler, who, in wandering hither and thither upon the globe, places an occasional white stone at certain points to which memory turns with never-failing pleasure. Thus he recalls a sunrise over the silvery peaks of the grand Himalayan range; a thrilling view from the Mosque of Mahomet Ali at Cairo, localizing Biblical story; or a summer sunset-glow on the glassy mirror of the Yellowstone Lake.

Along the mountain side, east of the lake, are

ancient terraces, indented shorelines, and other evidences which clearly prove that, at no very remote geological period, the surface of this grand sheet of water was at least five or six hundred feet higher than it is at the present time. Nearly two hundred square miles of the Park are still covered by lakes.

As to the flora of the Yellowstone Park, seventy-five per cent. of the whole area seems to be covered by dense forests, the black fir being the most plentiful, often growing to three or four feet in diameter and a hundred and fifty feet in height. The white pine is the most graceful among the indigenous trees, and is always remarkable for its stately symmetrical beauty. The thick groves of balsam fir are particularly fine and fragrant, while the dwarf maples and willows are charming features as they mingle abundantly with larger and more pretentious trees. Wild flowers, Nature's bright mosaics, are found in great variety during the summer, though there is rarely a night in this neighborhood without frost, while the winters are truly arctic in temperature. The larkspur, columbine, harebell, lupin, and primrose abound, with occasional daisies and other blossoms. Yellow water-lilies, anchored by their fragile stems, profusely sprinkle and beautify the surface of the shady pools. Exquisite ferns, lichens, and velvety mosses delight the appreciative eye in many a sylvan nook which is only invaded by squirrels and song-birds.

Here, as in the valley of the Yosemite, it is

melancholy to see the track of devastating fires caused by the half-extinguished blaze left by careless camping parties. It is difficult to realize how intelligent people can be so wickedly reckless as to cause such destruction. Many a forest monarch stands bereft of every limb by the devouring flames, and large areas are entirely denuded of growth other than the shrubbery which springs up quickly after a sweeping fire in the woods, as though Nature desired to cover from sight the devastating footsteps of the Fire King. The grasses grow luxuriantly, especially alpine, timothy, and Kentucky blue grass.

There are many wild animals in the Park, such as elk, deer, antelope, big-horn sheep, foxes, buffalo, and what is called the California lion, a small but rather dangerous animal for the hunter to encounter. The buffalo is rarely seen in the West, and it is said is now only to be found wild in this Park. The streams and creeks also swarm with otter, beaver, and mink. These animals are all protected by law, visitors being only permitted to shoot such birds as they can cook and eat in their camps, together with any species of bear they may chance to fall in with; and there are several kinds of the latter animal to be found in the hills. At least this has been the case until lately; but stricter rules have been found necessary, and no visitors are now permitted to take firearms with them while remaining in the Park. The purpose of the government is to strictly preserve the game, the effect of which has already been to render the animals gathered here less shy of human approach, and to greatly increase their number.

So abundant are the evidences of grand volcanic action throughout the lake basin that it has been looked upon by scientists as the remains or centre of one enormous crater forty miles across! Dr. Hayden, the profound geologist, who was sent professionally by the government to report upon the Park, declares it to have been the former scene of volcanic activity as great as that of any part of this planet, a conclusion which the observer of to-day is quite ready to admit, inasmuch as the subsidence has yet left enough of the original forces to demonstrate the sleeping power which still lurks restlessly beneath the soil. We wonder, standing amid such remarkable surroundings, how many centuries have passed since the valley assumed its present shape. Everything is indicative of high antiquity, and it is probably rather thousands than hundreds of years since this volcanic centre was at its maximum power and activity. The valley has been partly excavated out of ancient crystalline rocks, partly out of later stratified formations, and partly from masses of lava that were poured forth during a succession of ages which make up the different epochs of the earth's long history.

The lowest level of the Park is about six thousand feet above the sea, and the average elevation, independent of mountains, is much over this estimate. It is very properly designated as the summit of the continent, and gives rise to three of the largest rivers in North America, namely: on the north side are the sources of the Yellowstone; on the west, three of the forks of the Missouri; and on the southwest are the sources of the Snake River, which flows into the Columbia, and thence to the distant Pacific Ocean.

If possible, before leaving the neighborhood, the visitor should ascend Mount Washburn, the highest point of observation within the great reservation, a feat easily accomplished on horseback. Such an excursion is particularly desirable since all the scenery of the Park is circumscribed while we are at the level of its springs, geysers, and lakes. The grand view from this elevation will repay all the time and effort expended in its accomplishment. Its height above the base is five thousand feet, its height above the sea five thousand more. A clear day is absolutely necessary for the proper enjoyment of such an excursion, in order to bring out fairly the panorama of forests, lakes, prairies, and mountains, decked by the golden glory of the sunshine. In some directions the vision reaches a hundred and fifty miles through space. Here, on the summit of Mount Washburn, we virtually stand upon the apex of the North American continent, if we except one or two of the sky-reaching peaks of the Territory of Alaska.

As we face the north, just before us lies the valley of the Yellowstone, and in the distance, looming far above its surroundings, is the tall

Emigrant Peak. To the eastward Index and Pilot peaks pierce the clouds, beyond which stretches away the Big Horn Range. In the west the summits of the Gallatin Mountains follow one another northward, while trending in the same direction, but farther towards the horizon, is the lofty Madison Range. We gaze until bewildered by peak after peak, mountain beyond mountain, range upon range, mingling with each other, all combining to form a glorious view embodying the indescribably grand characteristics of the Rocky Mountain system, the equal of which we may never again behold.

The tall range of mountains which girdle the Park are snow-covered all the year round, frigid, giant sentinels, which long proved a complete barrier to organized exploration, forming an amphitheatre of sublime and lonely scenery. The story of the discovery of this Wonderland is briefly told as follows: It seems that a gold-seeking prospector named Coulter made his way with infinite perseverance into the region in 1807, and after many hair-breadth escapes from Indians, wild beasts, poisonous waters, and starvation, finally succeeded in rejoining his comrades, whom he entertained with stories of what he had seen, which seemed to them so incredible that they believed him to be crazy. Afterwards, first one and then another adventurer found his way hither, and though each of them corroborated Coulter's story, they were by no means fully credited. But public attention and curiosity were thus aroused, leading the government to send Professor Hayden and a small exploring party to carefully examine the region. This enterprise not only corroborated the stories already made public, but greatly added to their volume and amazing detail.

It was found that the representations of Coulter and those who followed him, so far from exaggerating the wonders of the Yellowstone, in reality fell far below the truth.

During the year 1870 Governor Washburn, accompanied by a small body of United States cavalry, entered the Park by the valley of the Yellowstone, and thoroughly explored the cañons, the shores of the great lake, and the geyser region of Firehole River, together with the various interesting localities of which we have spoken. On returning he declared that the party had seen the greatest marvels to be found upon this continent, and that there was no other spot on the globe where there were crowded together so many natural wonders, combined with so much beauty and grandeur.

Finally Congress, foreseeing that the greed of speculators would lead them to monopolize this Wonderland for mercenary purposes, promptly took action in the matter, setting the region aside as a National Park and Reservation, for the benefit of the people at large forever, retaining the fee and control of the same in the name of the government.

Not many persons have ever attempted to traverse the Park in the winter season, but it has been done by a few hardy and adventurous people, who nearly perished in the attempt. Such individuals have reported that the raging snow-storms and blizzards which they encountered were on a scale quite equal to the other demonstrations and natural curiosities of the place. The trees in their neighborhood were beautifully gemmed with the frozen vapor of the geysers, and the heated springs seemed doubly active by the contrast between their temperature and that of the freezing atmosphere. It was only by camping at night upon the very brink of these boiling waters that life could be sustained, with the atmosphere at forty degrees below zero.

One who comes hither with preconceived ideas of the peculiar sights to be met with is sure to be disappointed, not in their want of strangeness, for the Park is overstocked with curiosities having no counterpart elsewhere, but the features are so thoroughly unique that his anticipations are transcended both in the quality and the quantity of the food for wonder which is spread out before him on every side.

CHAPTER V.

Westward Journey resumed. — Queen City of the Mountains. — Crossing the Rockies. — Butte City, the Great Mining Centre. — Montana. — The Red Men. — About the Aborigines. — The Cowboys of the West. — A Successful Hunter. — Emigrant Teams on the Prairies. — Immense Forests. — Puget Sound. — The Famous Stampede Tunnel. — Immigration.

AFTER a delightful, though brief, sojourn of ten days in the Yellowstone Park, realizing that twice that length of time might be profitably spent therein, we returned to Livingston, where the Northern Pacific Railroad was once more reached, and the westward journey promptly resumed. The Belt Range of mountains is soon crossed, at an elevation of over five thousand five hundred feet. A remarkable tunnel is also passed through, three thousand six hundred feet in length, from which the train emerges into a grand cañon, and soon arrives at the city of Bozeman. This place has a thrifty and intelligent population of over five thousand, and is notable for its rural and picturesque surroundings, in the fertile Gallatin Valley, which is encircled by majestic ranges of mountains, shrouded in " white, cold, virgin snow." Having passed the point where the Madison and Jefferson rivers unite to form the headwaters of that great river, the Missouri, whence it starts upon its long and winding course of over four thousand miles towards the Mexican Gulf, we arrive presently at Helena, the interesting capital of Montana. This is called the "Queen City of the Mountains," and is famous as a great and successful mining centre, the present population of which is about twenty thousand. It is said to be the richest city of its size in the United States, an assertion which we have good reasons for believing to be correct. The vast mineral region surrounding Helena is unsurpassed anywhere for the number and richness of its gold and silver-bearing lodes, having within an area of twenty-five miles over three thousand such natural deposits, the ownership of which is duly recorded, and many of which are being profitably worked. The city is lighted by a system of electric lamps, and has an excellent water-supply from inexhaustible mountain streams.

We were told an authentic story illustrating the richness of the soil in and about Helena, as a gold-bearing earth, which we repeat in brief.

It seems that a resident was digging a cellar on which to place a foundation for a new dwelling-house, when a passing stranger asked permission to remove the pile of earth that was being thrown out of the excavation, agreeing to return one half of whatever value he could get from the same, after washing and submitting it to the usual treatment by which gold is extracted. Permission was granted, and the earth was soon removed. The citizen thought no more about the matter. After a couple of weeks, however, the stranger returned and handed the proprietor of the ground thirteen hundred dollars as his half of the proceeds realized from the dirt casually thrown out upon the roadway in digging his cellar.

Between Helena and Garrison the main range of the Rocky Mountains is crossed, and at an elevation of five thousand five hundred and forty feet the cars enter what is called the Mullan Tunnel. This dismal and remarkable excavation is nearly four thousand feet long. From it the western-bound traveler finally emerges on the Pacific slope, passing through the beautiful valley of the Little Blackfoot.

The region through which we were traveling stretches from Lake Superior to Puget Sound, on the Pacific coast, and spreads out for many miles on either side of the Northern Pacific Railroad, known as the "Northern Pacific Country." No portion of the United Sates offers more favorable opportunities for settlement, and in no other section is there as much desirable government land still open to preëmption, presenting such a variety of surface, richness of soil, and wealth of natural productions. Intelligent emigrants are rapidly appropriating the land of this very attractive region, but there is still enough and to spare. Europe may continue to send us her surplus population for fifty years to come at the same rate she has done for the past half century, and there will still be room enough in the great West and Northwest to accommodate them.

As we left the main track of the Northern Pacific Railroad at Livingston to visit the Yellowstone Park, so at Garrison we again take a branch road to Butte City, situated fifty-five miles southward, and which is admitted to be the greatest mining city of the American continent. Here, on the western slope of the main range of the Rocky Mountains, stands the "Silver City," as it is generally called, though one of its main features is its copper product, which rivals that of the Lake Superior district in quantity and quality, giving employment to the most extensive smelting works in the world. There are thirty thousand inhabitants in Butte, and it is rapidly growing in territory and population. Its citizens seem to be far above the average of our frontier settlers in intelligence and thrift. The Blue Bird silver mine is perhaps the richest in this locality, yielding every twelve months a million and a half of dollars in bullion; while the Moulton, Alice, and Lexington mines each produce a million dollars or more in silver yearly. There are several other rich mines, among them the Anaconda copper mine, which gives an aggregate each year larger in value than any we have named. The Parrott Copper Company, also the Montana and Boston Copper Company, each show an annual output of metal valued at a million of dollars. In place of there being any falling off in these large amounts, all of the mines are increasing their productiveness monthly by means of improved processes and enlarged mechanical facilities. But we have gone sufficiently into detail to prove the assertion already made, that Butte City is the greatest mining town on the continent. Eight tenths of its population is connected, either directly or indirectly, with mining.

"It would seem that the United States form the richest mineral country on the globe," said an English fellow-traveler to whom these facts were being explained by an intelligent resident.

"That has long been admitted," said the American.

"And what country comes next?" asked the Englishman.

"Australia," was the reply. "But the United States," continued the American, "have another and superior source of wealth exceeding that of all other lands, namely, their agricultural capacity. There are here millions upon millions of acres, richer than the valley of the Nile, which are still virgin soil untouched by the plow or harrow."

Not mining, but agriculture forms the great and lasting wealth of our broad and fertile Western States, rich though they be in mineral deposits, especially of gold and silver.

Before proceeding further on our journey, let us pause for a moment to consider the magnitude of this imperial State of Montana, which measures over five hundred miles from east to west, and which is three hundred miles from north to south, containing one hundred and forty-four thousand square miles. This makes it larger in surface than the States of New Hampshire, Vermont, Massachusetts, Connecticut, New Jersey, Maryland, Ohio, and Indiana combined. With its vast stores of mineral wealth and many other advantages, who will venture to predict its future possibilities? It would be difficult to exaggerate them. The precious metals mined in the State during the last year gave a total value of over forty million of dollars, which was an increase of six million over that of the preceding year. Between forty and fifty million dollars in value is anticipated as the result of the local mining enterprise for the current twelve months, and yet we consider this to be the second, not the first, interest of Montana; agriculture take the precedence.

Returning to Garrison, after a couple of days passed at Butte City examining its extremely interesting system of mining for the precious metals, we once more resume our western journey.

Along the less populous portions of the route groups of dirty, but picturesque looking Indians are seen lounging about, wrapped in fiery red blankets. These belong to various native tribes, such as the Sioux, Blackfeet, Cheyennes, and Arapahoes. Bucks, squaws, and papooses gather about the small railroad stations, partly from curiosity, and partly because they have nothing else to do; but they are ever ready to sell trifles of their own rude manufacture to travelers as souvenirs, also gladly receiving donations of tobacco or small silver coins. The men are fat, lazy, and useless, scorning even the semblance of working for a livelihood, leaving the squaws to do the trading with travelers. These are "wards" of our government, who receive regular annuities of money and subsistence, including flour, beef, blankets, and so on. Support is thus insured to them so long as they live, and no American Indian was ever known to work for himself, or any one else, unless driven to it by absolute necessity.

When the author first crossed these plains, nearly thirty years ago, before there was any transcontinental railroad, the Indian tribes were very different people from what we find them to-day. The men were thin in flesh, wiry, active, and constantly on the alert. They were ever ready for bloodshed and robbery when they could be perpetrated without much danger to themselves. Contact with civilization has changed all this. They have become fat and lazy. They have borrowed the white man's vices, but have ignored his virtues. When not fighting with the pale faces, the tribes were, thirty and forty years ago, incessantly at war with each other, thus actively promoting the fate which surely awaited them as a people. Their pride, even to-day, is to display at their belts not only the scalps of white men and women taken in belligerent times, but also the scalps of hostile tribes of their own race.

We believe most sincerely in fulfilling all treaty obligations between our government and the Indians, to the very letter of the contract, nor have we any doubt that our official agents have often been unfaithful in the performance of their duties;

but when we attempt to create saints and martyrs out of the Red Men, we are certainly forcing the canonizing principle. They are entitled to as much consideration as the whites, but they are not entitled to more. They are crafty and cruel by nature; this is, perhaps, not their fault, but it is their misfortune. Nothing is really gained in our fine-spun moral theories by attempting to deceive ourselves or others. The plain truth is the best.

A little way from the railroad station on the open prairie the camps of these aborigines may often be seen, consisting of a few rude buffalo hides or canvas tents, while a score of rough looking ponies are grazing hard by, tethered to stakes driven into the soil. Here and there in front of a tent an iron kettle, in which a savory compound of meat and vegetables is simmering, hangs upon a tripod above a low fire built on the ground, presided over by some ancient squaw, all very much like a gypsy camp by the roadside in far off Granada.

The male aborigines wear semi-civilized clothing made of dressed deerskins, and woolen goods indiscriminately mixed; their long coarse black hair, decked with eagle's feathers, hangs about their necks and faces, the latter often smeared with yellow ochre. Now and then a touch of manliness is seen in the bearing and facial expression of the bucks; but the larger number are debauched and degraded specimens of humanity, who impress the stranger with some curiosity, but with very little interest. Like the gypsies of Spain, they are incorrigible nomads, detesting the ordinary conventionalities of civilized life. The Indian women are clad in leather leggings, blue woolen skirts and waists, having striped blankets gathered loosely over their shoulders. No one can truthfully ascribe the virtue of cleanliness to these squaws. The papooses are strapped in flat baskets to the mothers' backs, being swathed, arms, legs, and body, like an Egyptian mummy, and are as silent even as those dried-up remains of humanity. Whoever heard an Indian baby cry? The mothers seemed to be kind to the little creatures, whose faces, like those of the Eskimo babies, are so fat that they can hardly open their eyes.

We are sure to see about these railroad stations in the far West an occasional "cowboy," clad in his fanciful leather suit cut after the Mexican style, wearing heavy spurs, and carrying a ready revolver in his belt. His long hair is covered by a broad felt sombrero, and he wears a high-colored handkerchief tied loosely about his neck. He enjoys robust health, is sinewy, clear-eyed, and intelligent in every feature, leading an active, open-air life as a herdsman, and being ever ready for an Indian fight or a generous act of self-abnegation in behalf of a comrade. He will not object on an occasion to join a lynching-party who happen to have in hand some horse-thief or a murderous scoundrel who has long successfully defied the laws. These cowboys are splendid horsemen, sitting their high-pommeled Mexican saddles like the Arabs. They are oftentimes educated young men, belonging to respectable Eastern families, seeking a brief experience of this wild, exposed life, simply from a love of independence and adventure. They are chivalric, and nearly always to be found on the side of justice, however quick they may be in the use of the revolver. Their life is spent amid associations, and in regions, where the slow process of the law does not meet the exigencies constantly occurring. The reader may be assured that they are nevertheless governed by a sense of "wild justice," in which an element of real equity predominates. To realize the skill which they acquire, one must see half a dozen of them join together in " rounding up " a herd of several hundred cattle, or wild horses, scattered and feeding on the prairie, and from the herds collect and sort out the animals belonging to different owners, all being distinctly branded with hot irons when brought from Texas or elsewhere. In doing this it is often necessary to lasso and throw an animal while the operator is himself in the saddle and his horse at full gallop. No equestrian feats of the ring equal their daily performances, and no Indian of the prairies can compare with them for daring and successful horsemanship. Indeed, an Indian is hardly the equal of a white man in anything, not even in endurance. "An intelligent white man can beat any Indian, even at his own game," says Buffalo Bill. Each one of the aborigines has his pony, and some have two or three, but they are as a rule of a poor breed, overworked and underfed. They are never housed, never supplied with grain, but subsist solely upon the coarse bunch grass of the prairie. The poor, uncared-for animals which are seen as described about the natives' encampments tell their own doleful story. The Indian ponies and the squaws are alike always abused.

As we cross these plains straggling emigrant teams are often seen, called " prairie schooners." The wagons as a rule are much the worse for wear, being surmounted by a rude canvas covering, dark and mildewed, under which a wife and four or five children are generally domiciled. A few domestic utensils are carried in, or hung upon the body of, the vehicle, — a tin dipper here, a water-pail there, a frying-pan in one place, and an iron kettle in another. These wagons are usually drawn by a couple of sorry-looking horses, and sometimes by a yoke of oxen. Beside the team trudges the father and husband, the typical pioneer farmer, hardy, independent, self-reliant, bound west to find means of support for himself and brood. Many such are seen as we glide swiftly over the iron rails, causing us to realize how steadily the stream of humanity flows westward, spreading itself over the virgin soil of the new States and Territories, and producing a growth in population no less legitimate than it is rapid. These pioneers are almost invariably farmers, and by adhering to their calling are sure to make at least a comfortable living.

While stopping at a watering-place in the early morning, the picturesque figure of a hunter was seen with rifle in hand. Over his shoulder hung the body of an antelope, while some smaller game was secured to his leathern belt. He had just captured these in the wild brown hills which border the plateau where our train had stopped. Cooper's Leather-Stocking Tales were instantly suggested to the mind of the observer, as he watched the careless, graceful attitude and bearing of the rugged frontiersman, whose entire unconsciousness of the unique figure which he presented was especially noticeable.

After traveling more than five hundred miles in Montana, which is surpassed in size only by Alaska and Dakota, we enter northern Idaho, attractive for its wild and picturesque scenery, — a territory of mountains, valleys, rivers, lakes, and prairies combined, second only to Montana in its mineral wealth, and possessing also some of the choicest agricultural districts in the great West, where Nature herself freely bestows the best of irrigation in uniform and abundant rains. While traveling in Idaho we find that the route passes through a magnificent forest region, where the trees measure from six to ten feet in diameter, and are of colossal height, such growing timber as would challenge comment in any part of the world, consisting mostly of white pine, cedar, and hemlock.

We soon cross into the State of Washington, its northern boundary being British Columbia and its southern boundary Oregon, from which it is separated for more than a hundred miles of its length by the Columbia River. Its form is that of a parallelogram, fronting upon the Pacific Ocean for about two hundred and fifty miles, and having a length from east to west of over three hundred and sixty miles. This State has immense agricultural areas, as well as being rich in coal, iron, and timber. We pause at Spokane Falls for a day and night of rest. It is on the direct line of the Northern Pacific Railroad, and is the principal city of eastern Washington, having the largest and best water-power on the Pacific slope. Government engineers report the water fall here to exceed two hundred thousand horse-power, a small portion only of which is yet improved, and that as a motor for large grain and flouring mills. Here we find a thrifty business community numbering over twelve thousand, the streets traversed by a horse railroad, and the place having electric lights, gas and public water works, with a Methodist and a Catholic college. It commands the trade of what is termed the Big Bend country and the Palouse district, and is the fitting-out place for the thousands of miners engaged in Cœur d'Alene County. In spite of the late disastrous fire which she has experienced, Spokane, like Seattle, will rapidly rise from her ashes. Official reports show that over nine million acres of this State are particularly adapted to the raising of wheat. Our route, after a brief rest at Spokane Falls, lies through Palouse County, where this cereal is raised in quantities proportionately larger than even in Dakota, and at a considerably less cost. Thirty-five to forty bushels of wheat to the acre is considered a royal yield in Dakota and the best localities elsewhere, but here fifty bushels to the acre are pretty sure to reward the cultivator, and even this large amount is sometimes exceeded. One enthusiastic observer and writer declares that Palouse County is destined to destroy wheat-growing in India by virtue of its immense crops, its favorable seasons, its economy of production, and its proximity to the seaboard.

In the western part of the State, on Puget Sound, the lumber business is the most important industry, giving profitable employment to thousands of people. The productive capacity of the several sawmills on the sound is placed at two million feet per day, and all are in active operation. A new one of large proportions was also observed to be in course of construction. The forests which produce the crude material are practically inexhaustible. The pines are of great size, ranging from eight to twelve feet in diameter, and from two hundred to two hundred and eighty feet in height. No trees upon this continent, except the giant conifers of the Yosemite, surpass these in magnitude. United States surveyors have declared, in their printed reports, that this State contains the finest body of timber in the world, and that its forests cover an area larger than the entire State of Maine.

The most productive hop districts that are

known anywhere are to be found in the broad valleys of this State, where hop-growing has become a great and increasing industry, yielding remarkable profits upon the money invested and the labor required to market the crop. The course of the railroad is lined with these gorgeous fields of bloom, hanging on poles fifteen feet in height, planted with mathematical regularity. Large fruit orchards of apples, pears, peaches, cherries, and other varieties are seen flourishing here; and residents speak confidently of fruit raising as being one of the most promising future industries of this region, together with the canning and preserving of the fruits for use in Eastern markets. We are reminded, in this connection, that the United States crop reports also represent Washington as producing more bushels of wheat to the acre than any other State or Territory within the national domain. This grand region of the far northwestern portion of our country is three hundred miles long, from east to west, and two hundred and forty miles from north to south, giving it an area in round numbers of seventy thousand square miles. That is to say, it is nearly as large as the States of New York and Pennsylvania combined.

The immigration pouring into the new State of Washington is simply enormous, its aggregate for the year 1889 being estimated at thirty-five thousand persons, the majority of whom come hither for agricultural purposes, and to establish permanent homes. One train observed by the author consisted of nine second-class cars filled entirely with Scandinavians, that is, people from Norway and Sweden, presenting an appearance of more than average sturdiness and intelligence.

As the Pacific coast is approached we come to the famous Stampede Tunnel, which is nearly ten thousand feet long, and, with the exception of the Hoosac Tunnel in Massachusetts, the longest in America. On emerging from the Stampede Tunnel the traveler gets his first view of Mount Tacoma, rising in perpendicular height to nearly three miles, the summit robed in dazzling whiteness throughout the entire year.

CHAPTER VI.

,Mount Tacoma. — Terminus of the Northern Pacific Railroad. — Great Inland Sea. — City of Tacoma and its Marvelous Growth. — Coal Measures. — The Modoc Indians. — Embarking for Alaska. — The Rapidly Growing City of Seattle. — Tacoma with its Fifteen Glaciers. — Something about Port Townsend. — A Chance for Members of Alpine Clubs.

THE city of Tacoma takes its name from the grand towering mountain, so massive and symmetrical, in sight of which it is situated. We cannot but regret that the newly formed State did not assume the name also.

This is the western terminus of the Northern Pacific Railroad, and is destined to become a great commercial port in the near future, being situated so advantageously at the head of the sound, less than two hundred miles from the Pacific Ocean. Its well-arranged system of wharves is already a mile and a half long, while there is a sufficient depth of water in any part of the sound to admit of safely mooring the largest ships. The reports of the United States Coast Survey describe Puget Sound as having sixteen hundred miles of shore line, and a surface of two thousand square miles, thus forming a grand inland sea, smooth, serene, and still, often appropriately spoken of as the Mediterranean of the North Pacific. It is indented with many bays, harbors, and inlets, and receives into its bosom the waters of numerous streams and tributaries, all of which are more or less navigable, and upon whose banks are established the homes of many hundred thrifty farmers.

History shows that long ago, before any Pilgrims landed at Plymouth, Spanish voyagers planted colonies on Puget Sound. From them the Indians of these shores learned to grow crops of cereals, though according to the ingenious Ignatius Donnelly's "Atlantis" they brought the art from a lost continent. Puget Sound may be described as an arm of the Pacific which, running through the Strait of Fuca, extends for a hundred miles, more or less, southward into the State of Washington. Nothing can exceed the beauty of these deep, calm waters, or their excellence for the purpose of navigation; not a shoal exists either in the strait or the sound that can interfere with the progress of the largest ironclad. A ship's side would strike the shore before her keel would touch the bottom. Storms do not trouble these waters; such as are frequently encountered in narrow seas, like the Straits of Magellan, and heavy snow-storms are unknown. The entire expanse is deep, clear, and placid.

Tacoma has about thirty thousand inhabitants to-day; in 1880 it had seven hundred and twenty! The assessed valuation eight years ago was half a million dollars. It is now over sixteen million dollars, and this aggregate does not quite represent the rapid increase of real estate. Here, months have witnessed more growth and progress in permanent business wealth and value of property than years in the history of our Eastern cities. At this writing there is being built a large and architecturally grand opera house of stone and brick which will cost quarter of a million dollars, besides which the author counted over forty stone and brick business edifices in course of construction, and nearly a hundred two and three story frame-houses for dwelling purposes, of handsome modern architectural designs. Away from the business centre of the city the residences are universally beautiful, with well-kept lawns of exquisite green, and small charming flower gardens fragrant with roses, syringas, and honeysuckles, mingling with pansies, geraniums, verbenas, and forget-me-nots. It is astonishing what an air of leisure and refinement is imparted to these dwellings by this means,—an air of retirement and culture, amid all the surrounding bustle and rush of business interests.

The city claims an ocean commerce surpassed in volume by no other port on the Pacific except San Francisco. Its substantial and well-arranged brick blocks, of both dwellings and storehouses, lining the broad avenues, are suggestive of permanence and commercial importance, while a general appearance of thrift prevails in all of the surroundings. Pacific Avenue is noticeably a fine thoroughfare,— the principal one of the town. The place seems to be thoroughly alive, and especially in the vicinity of the shipping. The author counted fifteen ocean steamers in the harbor, and there were at the same time as many large sailing vessels lying at the wharves loading with lumber, wheat, coal, and other merchandise, exhibiting a degree of commercial energy hardly to be expected of so comparatively small a community. We were informed that four fifths of the citizens were Americans by birth, drawn mostly from the educated and energetic classes of the United States, forming a community of much more than average intelligence. Young America, backed by capital, is the element which has made the place what it is. It was a surprise to find a hotel so large and well appointed in this city as the "Tacoma" proved to be; a five-story stone and brick house, of pleasing architectural effect, and having ample accommodations for three hundred guests. It stands upon rising ground overlooking the extensive bay. The view from its broad piazzas is something to be remembered.

Across Commencement Bay is a point of well-wooded land, called "Indian Reservation," where our government located what remains of the Modoc tribe who so long resisted the advance of the whites towards the Pacific shore. These former belligerents are peaceable enough now, fully realizing their own interests.

Statistics show that there is shipped from Tacoma, on an average, a thousand tons of native coal per day, mostly to San Francisco and some other Pacific ports. A large portion of this coal comes from valuable measures belonging to the

Northern Pacific Railroad Company, situated thirty or forty miles from Tacoma, and some from the Roslyn mines farther away. The Wilkinson and Carbonado mines form the principal source of supply for shipment, and the Roslyn for use on the railroad. These last are thirty-five thousand acres in extent. One of the many veins of the Roslyn coal deposit is estimated to contain three hundred million tons of coal, conveniently situated for transportation on the line of the Northern Pacific Railroad.

The great Tacoma sawmill does a very large and successful business, finding its motor in a steam engine of fourteen hundred horse-power, and having over seven hundred men on its pay-roll. This number includes mill-hands, dock-men, choppers, and watermen, the latter being the hands who bring the logs by rafts from different parts of the sound. There are a dozen other sawmills in and about the city. The lumber business of this region is fast assuming gigantic proportions, shipments being regularly made to China, Japan, Australia, and even to Atlantic ports. A whole fleet of merchantmen were waiting their turn to take in cargo while we were there. We believe that Tacoma will ere long become the second city on the Pacific coast, and perhaps eventually a rival to San Francisco. Its abundance of coal, iron, and lumber, added to its variety of fish and immense agricultural products, are sufficient to support a city twice as large as the capital of California.

One sturdy gang of men, who are bringing in

a large raft of logs, attracts our attention by their similarity of dress and general appearance, as well as by their dark skins and well-developed forms. On inquiry we learn that they are native Indians of the Haida tribe, who come down from the north to work through a part of the season as lumbermen, at liberal wages. They are accustomed to perilous voyages while seeking the whale and fishing for halibut in deep waters, commanding good wages, as being equal to any white laborers obtainable.

We embark at Tacoma for Alaska in a large and well-appointed steamer belonging to the Pacific Coast Steamship Company, heading due north.

The first place of importance at which we stop is the city of Seattle, the oldest American settlement on the sound, and now having a busy commercial population of nearly thirty thousand. It has an admirable harbor, deep, ample in size, and circular in form; the commercial facilities could hardly be improved. Here again are large substantial brick and stone blocks, schools, churches, and various public and private edifices of architectural excellence. Enterprise and wealth are conspicuous, while the neighboring scenery is grand and attractive. To the east of the city, scarcely a mile away, is situated a very beautiful body of water, deep and pure, known as Lake Washington, twenty miles long by an average of three in width, and from which the citizens have a never-failing supply of the best of water. The lake has an area of over sixty square miles, and is surrounded by hills covered with a noble forest-growth of fir, spruce, and cedar. Seattle has four large public schools averaging six hundred pupils each, and a university to which there are seven professors attached, with a regular attendance of two hundred students.

Among the great natural resources of this region there is included sixty thousand acres of coal fields within a radius of thirty miles of Seattle. These coal fields are connected with the city by railways. Tacoma and Seattle are also joined by rail, besides two daily lines of steamboats.

Great is the rivalry existing between the people here and those of Tacoma, but there is certainly room enough for both; and, notwithstanding the destructive fire which lately occurred at Seattle, it is prospering wonderfully. About four miles distant from the centre of business is situated one of the largest steel manufactories in this country, the immediate locality being known as Moss Bay. Here timber, water, coal, and mineral are close at hand to further the object of this mammoth establishment, which, when in full operation, will give employment to five thousand men. Real estate speculation is the present rage at Seattle, based on the idea that it is to be *the* port of Puget Sound.

Between the city and hoary-headed Mount Tacoma is one of the finest hop-growing valleys extant. It has enriched its dwellers by this industry, and more hops are being planted each succeeding year, increasing the quantity exported

by some twenty-five per cent. annually. It may
be doubted if the earth produces a more beautiful sight in the form of an annual crop of vegetation than that afforded by a hop-field, say of
forty acres, when in full bloom. We were told
that the land of King County, of which Seattle is
the capital, is marvelous in fertility, especially
in the valleys, often producing four tons of hay
to the acre; three thousand pounds of hops, or
six hundred bushels of potatoes, or one hundred
bushels of oats to the acre are common. It must
be remembered also that while there is plenty of
land to be had of government or the Northern Pacific Railroad Company at singularly low rates,
transportation in all directions by land or water is
ample and convenient, a desideratum by no means
to be found everywhere.

From the deck of the steamer, as we sail northward, the irregular-formed, but well-wooded shore
is seen to be dotted with hamlets, sawmills, farms,
and hop-fields, all forming a pleasing foreground
to the remarkable scenery of land and water presided over by the snow-crowned peak of Mount
Tacoma, which looms fourteen thousand feet and
more skyward in its grandeur and loneliness. How
awful must be the stillness which pervades those
heights! As we view it, the snow-line commences at about six thousand feet from the base,
above which there are eight thousand feet more,
ice-topped and glacier-bound, where the snow
and ice rest in endless sleep. There are embraced
within the capacious bosom of Tacoma fifteen

glaciers, three of which, by liberal road-making and engineering, have been rendered accessible to visitors, and a few persistent mountain climbers come hither every year to witness glacial scenery finer than can be found in Europe. Persons who have traveled in Japan will be struck by the strong resemblance of this Alpine Titan to the famous volcano of Fujiyama, whose snow-wreathed cone is seen by the stranger as he enters the harbor of Yokohama, though it is eighty miles away.

As we steam northward other peaks come into view, one after another, until the whole Cascade Range is visible, half a hundred and more in number.

The summit of Tacoma is not absolutely inaccessible. A dozen daring and hardy climbers have accomplished the ascent first and last; but it involves a degree of labor and the encountering of serious dangers which have thus far rendered it a task rarely achieved. Many have attempted to scale these lonely heights, and many have given up exhausted, glad to return alive from this perilous experience between earth and sky. Members of various Alpine clubs cross the Atlantic to climb inferior elevations. Let such Americans test their athletic capacity and indulge their ambition by overcoming the difficult ascent of Tacoma.

Port Townsend is finally reached, — the port of entry for Puget Sound district and the gateway of this great body of inland water. Tacoma, Seattle, and Port Townsend are all lively contestants for supremacy on Puget Sound. The business part of Port Townsend is situated at the base of a bluff which rises sixty feet above the sea level, upon the top of which the dwelling-houses have been erected, and where a marine hospital flies the national flag. To live in comfort here it would seem to be necessary for each family to possess a balloon, or that a big public lift should be established to take the inhabitants of the town from one part to the other. It is rapidly growing, — street grading and building of stores and dwelling-houses going on in its several sections. Vancouver named the place after his distinguished patron, the Marquis of Townshend. We were told that over two thousand vessels enter and clear at the United States custom-house here annually, besides which there are at least a thousand which pass in and out of the sound under coasting licenses, and are not included in this aggregate. The collections of the district average one thousand dollars for each working day of the year.

Port Townsend is nine hundred miles from San Francisco by sea, and thirty-five hundred miles, in round numbers, from Boston or New York. It is the first port from the Pacific Ocean, and the nearest one to British Columbia, besides being the natural outfitting port for Alaska. We were surprised to learn the extent of maritime business done here, and that in the number of American steam vessels engaged in foreign trade it stands foremost in all the United States. Its climate is said to be more like that of Italy than any other part of America. The place is certainly remarkable for salubrity and healthfulness, and is universally commended by persons who have had occasion to remain there for any considerable period. The view from the upper part of the town is very comprehensive, including Mount Baker on one side and the Olympic Range on the other, while the far-away silver cone of Mount Tacoma is also in full view. The busy waters of the sound are constantly changing in the view presented, various craft passing before the eye singly and in groups. Long lines of smoke trail after the steamers, whose turbulent wakes are crossed now and then by some dancing egg-shell canoe or a white-winged, graceful sailboat bending to the breeze.

Certain custom-house formalities having been duly complied with, we continued on our course, bearing more to the westward, crossing the Strait of Juan de Fuca, bound for Victoria, the capital of Vancouver Island and of British Columbia, at which interesting place we land for a brief sojourn. To the westward the port looks out through the Strait of Fuca to the Pacific, southward into Puget Sound, and eastward beyond the Gulf of Georgia to the mainland.

CHAPTER VII.

Victoria, Vancouver's Island. — Esquimalt. — Chinamen. — Remarkable Flora. — Suburbs of the Town. — Native Tribes. — Cossacks of the Sea. — Manners and Customs. — The Early Discoverer. — Sailing in the Inland Sea. — Excursionists. — Mount St. Elias. — Mount Fairweather. — A Mount Olympus. — Seymour Narrows. — Night on the Waters. — A Touch of the Pacific.

THE city of Victoria contains twelve thousand inhabitants, more or less, and is situated just seventy miles from the mainland; but beyond the fact that it is a naval station, commanding the entrance to the British possessions from the Pacific, we see nothing to conduce to the future growth of Victoria beyond that of any other place on the sound. The aspect is that of an old, steady-going, conservative town, undisturbed by the bustle, activity, and business life of such places as Tacoma and Seattle. Vancouver, on the opposite shore, being the terminus of the Canadian Pacific Railway, bids fair to soon exceed it in business importance, though it has to-day less than ten thousand inhabitants. The population of Victoria is highly cosmopolitan in its character, being of American, French, German, English, Spanish, and Chinese origin. Of the latter there are fully three thousand. They are the successful market-gardeners of Victoria, a position they fill in many of the

English colonies of the Pacific, also performing the public laundry work here, as we find them doing in so many other places. In the hotels they are employed as house-servants, cooks, and waiters. Yet every Chinaman who lands here, the same as in Australia and New Zealand, is compelled to pay a tax of fifty dollars entrance fee. The surprise is that such an arbitrary rule does not act as a bar to Asiatic immigration; but it certainly does not have that effect, while it yields quite a revenue to the local treasury. At most ports the importation or landing of Chinese women is forbidden, but some of the gayest representatives of the sex are to be seen in the streets of Victoria, with bare heads, having their intensely black hair, shining with grease, dressed in large puffs. The heavy Canton silks in which they are clothed indicate that they have plenty of money. They affect gaudy colors, and wear heavy jade ear-rings, with breastpins of the same stone set in gold. The lewd character of the Chinese women who leave their native land in search of foreign homes is so well known as to fully warrant the prohibition relative to their landing in American or British ports. The effort to exclude them is, however, not infrequently a failure, as with a trifling disguise male and female look so much alike as to deceive an ordinary observer. The Asiatics are up to all sorts of tricks to evade what they consider arbitrary laws.

Officially Victoria is English, but in population it is anything else rather than English. Until

1858 it was only a small trading station belonging to the Hudson Bay Company; but in that year the discovery of gold on the bar of the Fraser River and elsewhere in the vicinity caused a great influx of miners and prospectors, mostly from California, and it was this circumstance which gave the place a business start and large degree of importance. The houses are many of them built of stone and bricks, the gardens being also neatly inclosed. The streets are macadamized and kept in excellent order. The city is lighted by electric lamps placed on poles over a hundred feet high, and has many modern improvements designed to benefit the people at large, including large public buildings and a fine opera house.

The harbor of Victoria is small, and has only sufficient depth to accommodate vessels drawing eighteen feet of water; but near at hand is a second harbor, known as Esquimalt, with sufficient depth for all practical purposes. If quiet is an element of charm, then Victoria is charming; but we must add that it is also rather sleepy and tame. It might be centuries old, everything moving, as it does, in grooves. Business people get to their offices at about ten o'clock in the morning, and leave them by three in the afternoon. There is no evidence here of the fever of living, no symptom of the go-ahead spirit which actuates their Yankee neighbors across the sound.

Esquimalt is situated but three or four miles from Victoria, and is the headquarters of the English Pacific squadron, where two or three British men-of-war are nearly always to be seen in the harbor, and where there is also a very capacious dry-dock and a naval arsenal. At the time of our visit a couple of swift little torpedo-boats were exercising about the harbor and the sound. The well-wooded shore is dressed in "Lincoln green," far more tropical than boreal. The many pleasing residences are surrounded with pretty garden-plots, and flowers abound. We have rarely seen so handsome an array of cultivated roses as were found here. So equable is the climate that these flowers bloom all the year round. A macadamized road connects Esquimalt with Victoria, running between fragrant hedges, past charming cottages, and through delightful pine groves. We see here a flora of great variety and attractiveness, which could not exist in this latitude without an unusually high degree of temperature, accompanied with a great condensation of vapor and precipitation of rain. Victoria is admirably situated, with the sea on three sides and a background of high-rolling hills, and also enjoys an exceptionally good climate, almost entirely devoid of extremes.

The suburbs are thickly wooded, where palm-like fern-trees a dozen feet high, and in great abundance, recalled specimens of the same family, hardly more thrivingly developed, which the writer has seen in the islands of the South Pacific. The wild rose-bushes were overburdened with their wealth of fragrant bloom; we saw them in June, the favorite month of this queen of flowers. No wonder that Marchand, the old French voyager, when he found himself here on a soft June day, nearly a century ago, amid the annual carnival of flowers, compared these fields to the rose-colored and perfumed slopes of Bulgaria. If the reader should ever come to this charming spot in the far Northwest, it is the author's hope that he may see it beneath just such mellow summer sunshine as glows about us while we record these pleasant impressions in the queen-month of roses. Glutinously rich vines of various-colored honeysuckles were draped about the porticoes of the dwellings, whence they hung with a self-conscious grace, as though they realized how much beauty they imparted to the surroundings. The drone of bees and swift-winged humming-birds were not wanting, and the air was laden with their delicious perfume. The wild syringas, which in a profusion of snow-white blossoms lined the shaded roads here and there, were as fragrant as orange-blossoms, which, indeed, they much resemble. The air was also heavy with a dull, sweet smell of mingled blossoms, among which was the tall, graceful spirea with its cream-colored flowers, so thickly set as to hide the leaves and branches. The maple leaves are twice the usual size, and fruit-trees bend to the very ground with their wealth of pears, apples, and peaches. The alders, like the ferns, assume the size of trees, and cultivated flowers grow to astonishing proportions and beauty. The bark-shedding arbutus was noticeable for its peculiar habit, and its bare, salmon-colored trunk contrasting with its neighbors.

A portion of the site of Victoria is set aside as a reservation, and named Beacon Hill Park, containing choice trees and pleasant paths bordered with delicate shrubbery. But the whole place is park-like in its attractive picturesqueness. In the interior of the island there is said to be plenty of game, such as elk and red deer, foxes and beaver. These forests are dense and scarcely explored; sportsmen do not have to penetrate them far to find an abundance of game, so that in the open season venison is abundant and cheap in the town.

British Columbia, of which this city is the capital, embraces all that portion of North America lying north of the United States and west of the Rocky Mountains to the Alaska line. Its area is three hundred and forty thousand square miles, and it certainly possesses more intrinsic wealth than any other portion of the Dominion, except the eastern cities of Canada. It is but sparsely settled, and its natural resources are quite undeveloped.

The well-constructed roads in and about Victoria give it an advantage over most newly settled places, and the idea is worthy of all commendation. The seaward, or western shore of Vancouver, overlooking the North Pacific is very rocky, and is indented by frequent arms of the sea, like the fjords of Scandinavia, while the surface of the island is generally mountainous.

The Haidas and the Timplons are the two native tribes of Vancouver, who are represented to have once been very numerous, brave, and warlike.

Some of their canoes were eighty feet long, and most substantially constructed, being capable of carrying seventy-five fighting men, with their bows, arrows, spears, and shields of thick walrus hide. These war-boats were made from the trunk of a single tree, shaped and hollowed in fine nautical lines, so as to make them swift and buoyant, as well as quite safe in these inland waters. In these frail craft the natives were perfectly at home, and excited the admiration of the early navigators by the skill they displayed in managing them, so that Admiral Lütke named them the " Cossacks of the Sea."

But the Haidas, like the tribes of the Aleutian islands and the Alaska groups generally, have rapidly dwindled into insignificance — slowly fading away. People who subsist on fish and oil as staples can hardly be expected to evince much enterprise or industry. It cannot be denied, however, that as a race they appear much more intelligent and self-reliant than the aborigines of our Western States. Vincent Colyer, special Indian commissioner, says with regard to the natives of the southern part of Alaska and the Alexander Archipelago: " I do not hesitate to say that if three fourths of these Alaska Indians were landed in New York as coming from Europe, they would be selected as among the most intelligent of the many worthy emigrants who daily arrive at that port."

When these islands were first discovered by the whites, the native tribes occupying them were almost constantly at war one with another. The different tribes even to-day show no sympathy for each other, nor will they admit that they are of the same origin. Each has some theory of its exclusiveness and independence, all of which is a puzzle to ethnologists.

There seems never to have been any union of interest entertained among them. Before and after the advent of the Russians tribal wars raged among them incessantly. Blood was the only recognized atonement for offenses, and must be washed out by blood; thus vengeance was kept alive, and civil war was endless. Bancroft in his "Native Races of the Pacific" tells us that the Aleuts are still fond of pantomimic performances; of representing in dances their myths and their legends; of acting out a chase, one assuming the part of hunter, another of a bird or beast trying to escape the snare, now succeeding, now failing, until finally a captive bird is transformed into an attractive woman, who falls exhausted into the hunter's arms.

With well-screened foot-lights, verdant woodland surroundings, characters assumed by a trained ballet troupe, framed in the usual proscenium boxes, with orchestra in front, this would be a fitting entertainment for a first-class Boston or New York audience.

The Indians, or portions of the native race, seen in and about the streets of Victoria are of the most squalid character, dirty and unintelligent, being altogether repulsive to look upon.

The Indians of the west coast of the island are brought less in contact with the whites, and still keep up to a certain extent their native manners and customs, wearing fewer garments of civilization, and being satisfied with a single blanket as a covering during some portions of the year. They are fond of wearing curiously carved wooden masks at all their festivals, — some representing the head of a bear, some that of a huge bird, and others forming exaggerated human faces. There seems to be a spirit of caricature prevailing among them, as it does among the Chinese and Japanese.

These Vancouver aborigines have an original and extraordinary method of expressing their warm regard for each other, in isolated districts where they are quite by themselves. When they meet, instead of grasping hands or embracing, they bite each other's shoulders, and the scars thus produced are regarded with considerable satisfaction by the recipient. Their sacred rites are sanguinary, and their notions of religion are of a vague and incomprehensible kind. They believe in omens and sorcery, suffering as much from fear of supernatural evil as the most benighted African tribes. The west coast of Vancouver is nearly always bleak; the great waves of the North Pacific breaking upon it, even in quiet weather, with fierce grandeur, roaring sullenly among the rocks and caves.

The distant view from the eastern side of Vancouver is of a most charming character, embracing the blue Olympic range of mountains in the State of Washington, whose heads are turbaned with snow, while the lofty undulating peaks, taken *en masse*, resemble the fiercely agitated waves of the sea; a view which vividly recalled the Bernese Alps as seen from the city of Berne.

Vancouver is the largest island on the Pacific coast, and is well diversified with mountains, valleys, and long stretches of low pleasant shore. Its name commemorates that of one of the world's great explorers. Vancouver had served, previous to these notable explorations, as an officer under Captain Cook for two long and eventful voyages, and was thus well fitted for a discoverer and pioneer. He made a careful survey of Puget Sound with all of its channels, inlets, and bays, and wrote a faithful description of the coast of the mainland as well as of the islands. Though this was about a century ago, so faithfully did he perform his work that his charts are still regarded as good authority, though not absolutely perfect.

That practical seaman, in his sailing-ship, puts us to shame with all our science and steam facilities as regards surveys of this complicated region. The coast survey organization of the United States has done little more than to corroborate a portion of Vancouver's work. It is surprising that the government should neglect to properly explore and define by maps the islands, channels, and straits of the North Pacific coast. Notwithstanding our boasted enterprise, we are behind every power of Europe in these maritime matters.

The island of Vancouver has an area of eighteen

thousand square miles, and is therefore larger than Massachusetts, Rhode Island, Connecticut, and Delaware combined. It is only by these familiar comparisons that we can hope to convey clearly to the mind of the average reader such statistical facts, and cause them to be remembered.

Reference has been made to the favorable climate of Victoria. We should state that the maximum summer temperature is 84° Fah., and the minimum of the year is 22°.

From here our course lies in a northwest direction, leading through the broad Gulf of Georgia, which separates Vancouver from British Columbia. The magnificent ermine-clad head of Mount Baker is seen, for many hours, to the east of our course, looming far, far above the clouds, and radiating the glowing beauty of the sunset, which happened to be exceptionally fine at the close of our first day out from Victoria. The atmosphere, sea, and horizon were all the color of gold. The surface of the water was unbroken by a ripple, while it flashed in opaline variety the brilliant hues of the evening hour. The grand scenery which we encounter foreshadows the character of the voyage of a thousand miles, more or less, northward, to the locality of the great glaciers, forming a vast interior line of navigation unequaled elsewhere for bold shores, depth of water, numberless bays, and inviting harbors. The course is bordered for most of the distance with continuous forests, distinctly reflected in the placid surface of these straits and sounds. At times the passage, perhaps not more than a mile in width, is lined on either side with mountains of granite, whose dizzy heights are capped with snow, up whose precipitous sides spruce and pine trees struggle for a foothold, and clinging there thrive strangely upon food afforded by stones and atmospheric air. Occasionally we pass some deep, dark fjord, which pierces the mountains far inland, presenting mysterious and unexplored vistas. We come upon the island of San Juan, not long after leaving Victoria, which was for a considerable period a source of serious contention between England and America, the ownership being finally settled by arbitration, and awarded to us by the late Emperor of Germany. San Juan is remarkable for producing limestone in sufficient quantity to keep scores of lime-kilns occupied for a hundred years. The island was only important to us by its position, and as establishing certain boundary lines.

Now and again smoke is seen winding upwards from some rude but comfortable cabin on the shore, where a white settler and his Indian wife live in semi-civilized style. A rude garden patch adjoins the cabin, carpeted with thriving root crops, bordered by currant and gooseberry bushes, while numerous wooden frames are reared close by on which to dry salmon, cod, and halibut for winter use. Three or four half-breed children, with a marvelous wealth of hair, and clothed in a single garment reaching to the knees, watch us with open eyes and mouths as we glide along the smooth water-way. At last the father's attention is called to us by the exclamations of the papooses, and he waves us a salute with his slouchy fur cap. It is only a little spot on the lonely shore, but it is all the world to the squatter and his brood. One pauses mentally for an instant to contrast this type of lonely existence with the fierce and furious tide of life which exists in populous cities. Steamers, sailing craft, or native canoes have no storms to encounter here; the course is almost wholly sheltered, while coal or wood can be procured at nearly any place where the steamer chooses to stop. The fierce swell of the Pacific, so very near at hand, is completely warded off by the broad and beautiful islands of Vancouver, Queen Charlotte, Prince of Wales, Baranoff, and Chichagoff, which form a matchless panorama as they slowly pass, day after day, clad in thrifty verdure, before the eyes of the delighted voyager. Throughout so many hours of close observation one never wearies of the charming scene.

The trip between Victoria and Pyramid Harbor, in many of its features, recalls the voyage from Tromsöe, on the coast of Norway, to the North Cape, where the traveler beholds the grand phenomenon of the midnight sun, — passing over deep, still waters, winding through groups of lovely islands, covered with primeval forests and veined with minerals, amidst the grandest of Alpine scenery, where the nearer mountain peaks are clad in misty purple and those far away are wrapped in snow shrouds, where signs of human life are seldom seen, and the deep silence of the passage is broken only by the shrill cry of some wandering sea-bird. In both of these northern regions, situated in opposite hemispheres, grand mountains, volcanic peaks, and mammoth glaciers form the guiding landmarks. The glaciers of Alaska are not only many times as large as anything of the sort in Switzerland, but they have the added charm of the ever-changing beauties of the sea, thus altogether forming scenery of peculiar and incomparable grandeur. One often finds examples of the Scotch and Italian lakes repeated again and again on this inland voyage, where the delightful tranquillity of the waters so adds to the appearance of profound depth. It requires but little stretch of the imagination to believe one's self upon the Lake of Como or Lake Maggiore.

The enjoyment afforded to the intelligent tourist on this delightful route of travel is being more and more appreciated annually, as clearly evinced by the fact that over two thousand excursionists participated in the trips of steamers from Puget Sound to Sitka last year, by way of Glacier Bay and Pyramid Harbor, representing nearly every State in the Union, and also embracing many European travelers. "I thought it would be as cold as Greenland," said one of these tourists to us; "but after leaving Port Townsend I hardly once had occasion to wear my overcoat, night or day, during the whole of the fourteen days' summer voyage through Alaska's Inland Sea. The thermometer ranged between 68° and 78° during the whole trip, while the pleasant daylight never quite faded out of the sky."

Mount St. Elias, inexpressibly grand in its proportions, is probably the highest mountain in Alaska, and, indeed, is one of the half dozen loftiest peaks on the globe, reaching the remarkable height of nearly twenty thousand feet, according to the United States Coast Survey. It may fall short of, or it may exceed, this measurement by a few hundred feet. Owing to the low point to which the line of perpetual snow descends in this latitude, St. Elias is believed to present the greatest snow climb of all known mountains. Another notable peculiarity of this grand elevation is, like that of Tacoma, in its springing at once from the level of the Pacific Ocean, whereas most mountains, like those of Colorado, Norway, and Switzerland, say of twelve or fourteen thousand feet in height, rise from a plain already two or three thousand feet above sea level, detracting just so much from their effectiveness upon the eye, and from their apparent elevation. Vitus Behring, a Dane by birth and the discoverer of the strait which bears his name, first sighted this mountain on St. Elias' day, and so gave it the name which it bears. When the American whalemen on the coast saw the summit of Mount Fairweather from the sea, they felt sure that some days of fair weather would follow, hence we have the expressive name which is bestowed upon it. Mount St.

Elias, with its snow and ice mantle reaching nearly down to sea level, is higher than any elevation in Norway or Switzerland, rising from its base in pyramid form, straight, regular, and massive, to three times the height of our New England giant in the White Mountain range of New Hampshire, namely, Mount Washington. Only the Himalayas and the Andes exceed it in altitude. Eleven glaciers are known to come down from the south side of St. Elias, one of which, named Agassiz Glacier, is estimated to be twenty miles in width and fifty in length, covering an area of a thousand square miles!

Fairweather is situated about two hundred miles southeast of Mount St. Elias, its hoary head being often visible a hundred miles and more at sea; rising above the fogs and clouds, its summit is recognizable while all other land is far below the horizon. We were told that when the earthquake occurred at Sitka in 1847, this mountain emitted huge volumes of smoke and vapor. The force of volcanic action in Alaska is, however, evidently diminishing, though occasional slight shocks of earthquakes are experienced, especially on the outlying islands of the Aleutian group and near the mouth of Cook's Inlet.

Besides these loftiest mountains named, — "Rough quarries, rocks, and hills whose heads touch heaven," — Mount Cook, Mount Crillon, and Mount Wrangel should not be forgotten. Lieutenant H. T. Allen, U. S. A., makes the height of the latter exceed that of Mount St. Elias, but we think it very questionable. This officer's statement that Mount Wrangel is the birthplace of some of the largest glaciers known to exist seems much more likely to be correct. In this region, therefore, this far northwest territory of the United States, we find the highest elevations on the North American continent. The mountain ranges of California and Montana unite with the Rocky Mountains, and turning to the south and west form the Alaska Peninsula, finally disappearing in the North Pacific, except where a high peak appears now and then, raising its rocky crest above the sea, like a giant standing breast-high in the ocean, and thus they form the Aleutian chain of treeless islands, which stretch away westward towards the opposite continent. That these islands are all connected beneath the sea, from Attoo, the most distant, to where they join the Alaska Peninsula, is made manifest by the exhibition of volcanic sympathy. When one of the lofty summits emits smoke or fiery débris the others are similarly affected, or at least experience slight shocks of earthquake. So the several islands which form the Hawaiian group are believed to be joined below the ocean depths, and several, if not all, of the islands of the West Indies are considered to be similiarly connected.

This has been in some period, long ago, a very active volcanic region, as the lofty peaks, both among the Aleutian Islands and on the mainland, which emit more or less smoke and ashes, clearly testify; not only suggestive of the past, but significant of possible contingencies in the future. There are, in fact, according to the best authorities, sixty-one volcanic peaks in Alaska. One of the extinct volcanoes near Sitka, Mount Edgecombe, according to the Coast Pilot, has a dimension at the ancient crater of two thousand feet across, and an elevation of over three thousand feet above the sea. The depth of the crater is said to be three hundred feet. From the top, radiating downwards in singular regularity, are the deep red gorges scored by the burning lava in its fiery course, as thrown out of the crater less than a hundred years ago.

This is a Mount Olympus for the natives, about which many ancient myths are told by these imaginative aborigines.

For more than twenty-four hours after sailing from Victoria the irregular, kelp-fringed shore of Vancouver, which is three hundred miles long, is seen on our left, until presently the large, iron-bearing island of Texada, with its tall summit, appears on the right of our course. The magnetic ore found here in abundance is of such purity as to render it suitable for the manufacture of the highest grade of steel, and it is shipped to the furnaces at Seattle and elsewhere for this purpose.

It is found in pursuing the voyage northward that the fierce tide-way prevailing in some of the deep, narrow channels produces such turbulent rapids that steamers are obliged to wait for a favorable condition of the waters before attempting their passage, as the adverse current runs at the rate of nine miles an hour. This was especially the case in the Seymour Narrows, which is about nine hundred yards wide, and situated at no great distance from Nanaimo, in the Gulf of Georgia. It is a far more tumultuous water-way, at certain stages of the tide — which has a rise and fall of thirteen feet — than the famous Maelstrom on the coast of Norway. The latter is also caused by the power of the wind and tide, though it was long held as the mystery and terror of the ocean.

The author remembers in his school geography a crude woodcut, which depicted a ship being drawn by some mysterious power into a gaping vortex of the ocean, and already half submerged. It was intended to represent the terrible perils of passing too near the Maelstrom, off the Lofoden Islands. In after years he sailed quietly across this once dreaded spot in the North Sea, without experiencing even an extra lurch of the ship. Thus do the marvels and terrors of youth melt away. Travel and experience make great havoc in the wonderland of our credulity, and yet modern discovery outdoes in reality the miracles of the past.

A powerful steamer which attempted to pass through the Seymour Narrows at an unfavorable state of the water, last season, was unable to make way against the current, and came near being wrecked. By crowding on all steam she succeeded in holding her position until the waters subsided, though she made no headway for two hours. It was here that the United States steamer Saranac was lost a few years since, being caught at disadvantage in the seething waters, and forced upon the mid-channel rocks. Her hull now lies seventy fathoms below the surface of the sea. Since this event took place the United States ship Suwanee struck on an unknown rock farther north, and was also totally wrecked. Perhaps after a few more national vessels are lost in these channels our government will awaken from its lethargy, and have a proper survey made and reliable charts issued of this important coast and its intricate water-ways. A single vessel is now engaged in this survey, but half a dozen should be employed in Alaskan waters. Nanaimo is situated on the east side of Vancouver Island, seventy miles from Victoria, with which it is connected by railroad. It is a thrifty little town, mainly supported by the coal interest, though there are two or three manufacturing establishments. The extensive coal mines in its neighborhood are of great value, and are constantly worked. These coal deposits are of the bituminous sort, particularly well adapted for steamboat use, and are so situated as to facilitate the growing commerce of these islands. Many thousands of tons are shipped during the summer months to San Francisco. We are told that it cost the proprietors of these coal mines one dollar and a half a ton to place the product on board steamers, which on arriving at San Francisco fetches from twelve to fifteen dollars per ton.

There are five mines worked here, giving employment to some two thousand men, who receive two dollars and a half per day as laborers.

There is not a lighthouse upon any headland amid all of these meandering channels, though it must be admitted that navigation is rarely impeded for want of light in summer, as one can see to read common print at midnight upon the ship's deck without artificial aid any time during the traveling or excursion season of the year.

Now and again we look ahead inquiringly as we thread the labyrinth of islands and wonder how egress is possible from the many mountainous cliffs rising, sullen and frowning, directly in the steamer's course. The exit from this maze is quite invisible; but presently there is a swift turn of the wheel, the rudder promptly responds, and we gracefully round a projecting point into another lonely, far-reaching channel framed by granite peaks a thousand feet in height.

At night, when all but the watch were sleeping, how gaunt and weird stood forth those tall, black sentinel rocks, past which we were gliding so silently, while overhead was spread the broad firmament of space, dimly lighted by heaven's distant lamps! How suggestive the dark, mysterious shadows! how active the imagination! Was the atmosphere indeed peopled with the invisible spirits of bygone ages? Did the air-waves vibrate with the history of the long, long past, the unknown story of these silent fjords and deep water gorges? Is it only thousands, or tens of thousands, of years since the first human beings appeared and disappeared among these now wild, untrodden shores?

The inlets which are found at the head of the Gulf of Georgia, northeast of Vancouver Island, are miniature Norwegian fjords, deeper and darker than the sombre Saguenay; a hundred and eighty fathoms of line will not reach the bottom. They are from forty to sixty miles in length, with an average width of nearly two miles, being walled by abrupt mountains from four to seven thousand feet in height. A grand elevation, whose name has escaped us, stands eight thousand feet above the sea at the head of Butte Inlet, while Mount Alfred, at the head of Jarvis Inlet, is still higher. A remarkable feature of these elongated arms of the sea is their great depth, some of them measuring over three hundred fathoms. It is a popular idea that the phosphorescence of the sea is exhibited in its strongest effect in the tropics; but we have seen in the Gulf of Georgia, after sunset, so brilliant an illumination from this cause that it was only comparable to liquid fire, quite equal in intensity to anything the author has witnessed in the Indian Ocean or the Caribbean Sea. It is impossible to convey by the pen an idea of the novel splendor of the scene. A drop of this flame-like water, dipped from the sea in equatorial or Arctic waters and placed under the microscope is found to be teeming with the most curious living and active organisms. These myriads of tiny creatures are so minute that, were it not for the revelations of the microscope, we should not even know of their existence. Nor are these infinitesimal objects the smallest representatives of animal life; glasses of greater power will show still more diminutive creatures.

Persons who are accustomed to make sea-voyages do not forget to supply themselves with a good but inexpensive microscope, for use on shipboard. The abundant specimens of minute animal and vegetable life which the sea affords, form a source of instructive amusement by which many otherwise monotonous hours are pleasantly beguiled. A little familiarity with the instrument enables one to profitably entertain a whole ship's company with its powers.

In the region between Vancouver and Queen Charlotte Island we cross an open reach of the sea, and while the Pacific swell tosses us about after the usual erratic fashion of its unpacific waters, we observe a few ocean sights which serve pleasantly to vary the experience of the trip. A school of humpback whales put in an appearance, full of sport and frolic, in such extraordinary numbers that three or four are seen in the act of spouting all the while. In spots the sea is yellow, where its surface is covered for acres together with that animated food for other piscatory creatures, the jelly-fish. The shining, furry head of a sea-lion comes up to the surface now and again, gazing curiously at us with big, glassy eyes, and turning its face nimbly from side to side. A school of porpoises play about the hull of the steamer, leaping high out of the water and falling back again in graceful curves. The only shark we chanced to meet with on the entire voyage was observed in our wake just before entering Smith's Sound, south of Calvert Island. In this region the huge gona-bird was seen sailing slowly on the wing, recalling the albatross of the low latitudes in its long, lazy sweeps, as well as by its size and gracefulness. These bird-monarchs of the north measure eight feet from tip to tip, and glide with or against the wind on their broad, outspread pinions without the least visible muscular exertion, a mystery of motive power which is sure to challenge the observer's curiosity.

In the narrow passages the tall peaks, arched by the soft gray of the clouds and the clear blue of the sky, cast deep shadows where the water looked like pools of ink, whose blackness intensified the fact of their great but unknown depth.

The American whalers have never been accustomed to seek their big game in these immediate waters, preferring to attack the leviathans in lesser depths, such as the waters of Behring Sea, or farther north in the vicinity of the strait, between the frozen ocean and the North Pacific. There, if a whale dove after being struck by the harpoon, he was sure very soon to fetch up in the muddy bottom; but here, among the channels of the islands, he might dive, and dive again, to almost any depth, and unless great care was taken he was liable in his lightning-like velocity to carry down with him a whole boat's crew and all their belongings. Were it not that the whaling industry has gradually declined here, as it has done in all other sections of the globe, the possession of Alaska, with its great number of safe harbors, would be an invaluable boon to those of our countrymen engaged in that branch of commercial enterprise.

Inland sea travel is the perfection of steamboating, but the rapidly-changing landscape of these wild Alaskan shores, rimmed with sharp volcanic peaks, at last wearies the senses, and one is forced to seek a brief intermission by finding rest in sleep, only, however, to again renew the charm with greater zest on the morrow.

CHAPTER VIII.

Steamship Corona and her Passengers. — The New Eldorado. — The Greed for Gold. — Alaska the Synonym of Glacier Fields. — Vegetation of the Islands. — Aleutian Islands. — Attoo our most Westerly Possession. — Native Whalers. — Life on the Island of Attoo. — Unalaska — Kodiak, former Capital of Russian America. — The Greek Church. — Whence the Natives originally came.

OUR journey through that portion of Alaska known as the Inland Sea was made in the steamship Corona, Captain Carroll, a commander who has had long experience in these waters. His pleasure seemed to lie in the degree of enjoyment which he could afford his passengers, and the amount of information which he was enabled to impart to them. There were on board the Corona the members of a large excursion party conducted by Raymond & Whitcomb of Boston, numbering some eighty persons. We have rarely seen together a large party of ladies and gentlemen embracing so many cultured and agreeable persons. They had already occupied some weeks in a tour of Mexico and southern California. It was exceedingly pleasant to see the courtesy and consideration exercised among them towards each other, — amenities which go so far to lighten the inevitable inconveniences of travel, and to enhance its enjoyments. Oftentimes friendships are formed under such circumstances which continue through every exigency to the very end of life.

Having reached latitude 54° 40′ (the fifty-four forty or fight of 1862), we come to the boundary line between British Columbia and the United States, Dixon Entrance being on the left and Fort Tongas on the right. Here the far-reaching Portland Canal, or more properly channel, penetrates the mainland for a great distance, precisely like the Norwegian fjords, presenting, with its various arms, stupendous watery cañons, whence arise mountain precipices thousands of feet high on either side of the deep narrow course, their heads shrouded in perpetual snow. This channel, or fjord, runs nearly due north, and forms a boundary line to its head between the English and United States possessions.

Opposite and just south of Fort Tongas lies Fort Simpson, on British soil, and close at hand is Metla-katla, where that self-sacrificing missionary, Mr. Duncan, gathered and established a village of a thousand Christian residents from the various savage tribes of the vicinity. By his individual effort, with almost miraculous success, he raised from the lowest depths of barbarous life a law-abiding, religious, industrious, and self-supporting community, who justly considered him their moral and physical savior. Official persecution drove Mr. Duncan from Metla-katla to the nearest available American island, namely, Annetta, lying some sixty miles northward. Eight hundred of these aborigines whom he had reclaimed from savage life and its terrible practices have followed him with their families, freely abandoning all their property and improvements at Metla-katla, and are now struggling to create for themselves a new and permanent home under the United States.

The Senate committee, whose members lately visited Alaska, made a call at Annetta, and "found," as one of its members writes to the press, "the Indians living in an apparent condition of contentment, and engaged in almost all the pursuits of the whites. Their execution of artistic designs upon silver wrought by themselves into bracelets, rings, and all kinds of jewelry is marvelous. Baskets made in brilliant colors from stripped reeds constitute a beautiful and artistic employment of most of the women of the tribe. Their particular ambition is their anxiety to possess lands in severalty, or to have certain parcels set aside for them, that they may cultivate and hold in individual right. They ask that the whole of Gravine Island be given to their tribe. They found the state of the morals of the Indian women at Annetta; or, as they call it, New Metla-katla, far above the average of Indian women of this Territory. At Sitka the committee visited the habitations of the Indians, and learned much from personal intercourse as to their habits and needs. It was found that the companionship and virtue of the women is a matter of simply dollars and cents, and not difficult to negotiate for."

"The committee were surprised to observe such

an apparent freedom from rowdyism, quarrels, and disturbances of any character in any portion of the Territory, and remarked the entire absence of six-shooters about the person of a single individual, a feature always so prominent in the mining camps of the West."

Until Alaska — THE NEW ELDORADO — came into our possession, it was from the persistent and adventurous fur-traders that our knowledge of the country was almost solely obtained. To most of the public it was (and is still to many) scarcely more than a geographical expression, occupying an insignificant space on the extreme northwest portion of the maps of North America, without any regard being paid to the scale on which the other States and Territories of the country are delineated. The fact nevertheless stares us in the face, that Alaska is nearly as large as the whole of the United States lying east of the Mississippi River, or three times as large as France. Within the last twenty years greater intelligence has been shown, in part through missionaries, — self-sacrificing and devout men, — who have sought by their teachings to abolish the wild superstitions of the natives, together with their cruel rites of Shamanism. Organized companies of explorers, as well as enterprising miners and prospectors, have also liberally furnished us with general information relating to this great outlying province, which has been found to be so full of mineral wealth and future promise. But so vast is the Territory, so varied the climate, and so undeveloped are the means of access to its several parts, that our information as regards detail is still very meagre. There are not ten miles of roadway in all of Alaska outside of the island of Kodiak; or rather, we should say, the island just opposite Kodiak, namely, Wood Island, which has a road constructed completely round it, covering a dozen miles or thereabouts. The only road at Sitka is not over a mile and a half in length, and these two are the only ones in this vast Territory. Two objects of commercial gain, the profitable fur-trade and seeking for gold, have been the great agents of progress and development thus far in Alaska. In a like manner it was the greed for gold that first sent the Spaniards to Mexico and Peru; in pursuit of the lucrative fur-traffic the French and Britons opened the way for civilization in Canada. Here in Alaska it will not be philanthropy, — some of whose noblest exponents are upon the ground, — but self-interest; not government enterprise, but the seeking for precious metals, which will gradually unfold the great wealth and resources of this extensive province, whose area is greater than the thirteen original States of this Union. The hope of commercial gain has doubtless done nearly as much for the cause of truth and progress as the love of truth itself. The course of multitudes, guided by the natural instinct of selfishness, will be overruled by a higher power for the general good.

The very name of Alaska has to the popular ear a ring of glacier fields and snow-clad peaks, conveying a frigid impression of the climate quite contrary to fact. The most habitable portions of the country lie between 55° and 60° north, about the same latitude as that of Scotland and southern Scandinavia, but the area of this portion of Alaska is greater than that of both these countries combined. The name is derived from Alay-ck-sa, which was given to the mainland by the aborigines, and which signifies "great country." On the old maps it is very properly designated as Russian America, and so it really was until its transfer from the possession of that government to our own. It was at the request of Charles Sumner, whose able, eloquent, and consistent advocacy did so much towards its acquirement, that the aboriginal title of Alaska was adopted. The portion of the country which is at present visited by excursionists is the southeastern coast line and the archipelago of the Sitkan Islands or Alexander group. If one desires to reach the vast country and islands lying to the west and northwest, the proper way to do so is to sail direct from San Francisco for Unalaska and Kodiak. The last named island lies south of Cook's Inlet, one of the most remarkable volcanic regions in the Territory. Sitka is five hundred and fifty miles to the eastward of Kodiak. Cook's Inlet is well named, as the great discoverer sailed to its very head in 1778, being the first white man who ever did so, and, indeed, few have done it since. This was while he was prosecuting his vain search for a northwest passage around the continent of America. The finest and largest salmon which were ever known are taken in Cook's Inlet, reaching the weight of one hundred pounds in some instances, and measuring six feet in length. The island of Kodiak is also famous for its excellent and abundant salmon fisheries.

In 1874 a committee from the Icelandic residents of Wisconsin, aided by our government, made an excursion to Alaska to determine whether it would be advisable to recommend their people in Iceland to seek homes in and about Kodiak. The report of this committee, which consisted of three experienced and intelligent men, was published from the government printing-office in Washington, and from it we quote as follows:—

"Potatoes grow and do well, although the natives have not the slightest idea of how they should be cultivated, which goes to show they would thrive excellently if properly cared for. Cabbages, turnips, and the various garden vegetables have great success, and to judge from the soil and climate there is no reason why everything that succeeds in Scotland should not succeed at Kodiak. Pasture land is so excellent on the island, and the hay harvest so abundant, that our countrymen would here, just as in Iceland, make sheep breeding and cattle-raising their chief method of livelihood. The quality of the grass is such that the milk, the beef, and mutton must be excellent; and we had also an opportunity to try these at Kodiak."

The purpose of colonizing portions of Alaska with people from Iceland is being revived, and active measures to this end are now progressing. The people of that country are eager to avail themselves of such an opportunity. They are being gradually crowded out of their native land by the increased flow of volcanic matter over their plains and valleys. Alaska, while it affords them in certain portions, say the valley of the Yukon, a climate similar to their own, offers them also many advantages over the place of their nativity. It is authoritatively stated that over fifty thousand souls will gladly avail themselves of this chance to emigrate to Alaska, provided our government will aid them in the matter of transportation. At this writing, in the village of Afognak, on the island of Kodiak, with a population of three hundred natives, over one hundred acres of rich land is planted in potatoes and turnips, and has yielded annually a large crop of excellent vegetables for three or four consecutive years. If it were necessary we could point to several other successful agricultural developments in islands even less favorably situated than is the Kodiak group. Nevertheless, there are plenty of writers who assert that domestic vegetables will not grow in Alaska. One has no patience with such perversion of facts.

Miss Kate Field says in a late published article relative to Alaska: "In agriculture Alaska is not promising, but the country is by no means as impossible in this respect as it has been represented. 'There is not an acre of grain in the whole territory,' wrote Whymper. Because there was no grain grown, it by no means follows that grain cannot be grown in certain localities. Hundreds of acres of land near Wrangel can be drained and cultivated. The Indians on the neighboring islands raise tons of potatoes and turnips for their own consumption. Butter made for me by the Scotch housekeeper of Wrangel mission was a sweet boon, and proved that cows were a success in that region, and that dairies were a mere question of time."

The island of the Aleutian group situated the farthest seaward is named Attoo, and forms the most westerly point of the possessions of the United States. This island is situated about seven thousand five hundred miles in a straight line from the eastern coast of Maine, and is a little over three thousand miles west of San Francisco, making that city about the central point between the extreme east and west of this Union. It would be nearer, if one desired to reach England from Attoo, to continue his journey westward, rather than to travel east and cross the Atlantic. A few moments' examination of the globe or a good map of the world is especially desirable in this connection, and unless one is already familiar with this region will prove interesting and instructive. The Aleutian group, besides innumerable islets and rocks, contains over fifty islands exceeding three miles in length, seven of them being over forty miles long. Unimak, which is the largest, is over seventy miles long, with an average width of twenty.

It seems almost impossible to conceive of these islands having ever been densely populated, where human life is so sparsely represented to-day, and yet scientific investigation gives ample proof that in the far past every cove and bay echoed to the cry of the successful otter hunter, and the beaches now lined with numberless bidarkas or native canoes. The mummies which W. H. Dall brought hence may have been ten centuries old. This able investigator tells us of ruined villages and deserted hearths, to be found in almost any sheltered cove or favorably situated upland. A few strokes of the pick and the spade is sure to unearth arrow-heads, stone axes, and chipped implements of flint, or perhaps even the singularly proportioned bones of a now extinct human race. Bones have been exhumed on these islands which have puzzled scientists to account for.

When these islands were discovered by the Russians the inhabitants of Attoo were numerous, warlike, and brave, being well supplied with otter skins, and altogether were a self-reliant and thrifty tribe. Now the place contains but one small village, numbering about a hundred and twenty souls, situated on the south side of the island in a sheltered cove.

There are residents living upon Attoo to-day who have in their time witnessed two wrecks of Japanese vessels upon their shores; and who can say that Attoo was not originally peopled in this manner by Asiatics thousands of years ago? It was so late as 1861 that the last Japanese junk was stranded upon the island; three of the Japanese sailors surviving were ultimately sent home by way of Siberia overland.

The sea-otter has been driven from this immediate neighborhood by too vigorous and indiscriminate pursuit, but the sea-lion, various waterfowls, and plenty of cod, halibut, and salmon still abound among these lonely islands of the North Pacific. Occasionally a dead whale is stranded on the shore, which is considered a cause for great rejoicing, every part of the animal being utilized by the natives. No matter how putrid the flesh may be, it is eagerly eaten by these people, both raw and cooked. When a school of whales appears in sight of these shores, the natives go out in their frail boats, and with lances so prepared as to work into the vitals of the big creatures, they pierce them in the most vulnerable places, leaving the animal to die where it will, and trusting to the currents to carry the body where they can reach it. To their lances there are securely attached inflated sealskin buoys, which render diving a very laborious exertion to the whales, and which aid finally in securing the carcass. In this way, it is said, the natives get one whale out of fifteen or twenty which they succeed in harpooning. Whales, singular to say, are more esteemed as food by all the Alaskan shore tribes than any other product of the sea, or, in fact, any other sort of food. The securing of one is an event celebrated with limitless feasting and rejoicing. A New England whale-ship captain told the writer that he had seen these natives cut long strips of blubber from the body of a stranded whale, which had been so long dead that it was with difficulty he could breathe the atmosphere to leeward of the carcass, and chew upon the same with the greatest relish until it had entirely disappeared down their throats, the oil dripping all the while in small streams from the corners of their mouths. This is not a practice confined to the Aleuts, but extends throughout the several groups of islands, and is also a marked habit of the Eskimos proper, living both north and south of Behring Strait, and on the coast of the Polar Sea.

"The natives would rather have a dead whale drift ashore," says Mr. George Wardman, United States Treasury agent in Alaska, "than to own the best crop of the biggest farm in the United States. Dead whale is a great blessing in the Aleutian part of our Alaska possessions, and agricultural products are but little sought after or valued. The dead whale may be so putrid that the effluvia arising from it will blacken the white paint of a vessel lying one hundred yards distant, but, all the same, the whale is a blessing."

There is a variety store kept on Attoo by an agent of the Alaska Commercial Company, where the natives exchange their furs for tea, sugar, and hard biscuit, besides tobacco and a few fancy articles.

The mountains which surround the settlement are two or three thousand feet in height, "rock-ribbed and ancient as the sun," and are white with snow for a considerable portion of the year. These Aleutian Islands, bounded by wave-battered rocks, stretching far out in the Pacific towards Asia, have no trees, the soil not having sufficient depth to support them, but they are thickly covered with a low-growing, luxuriant vegetation in great variety. Between the mountains and the sea are many natural prairies, with a rich soil of vegetable mould suitable for domestic gardening. The wood consumed by the inhabitants as fuel is the product of drift-logs or trees reclaimed from the sea. On the breaking up of winter in the large islands at the northeast and on the mainland, the unsealing of the ice-bound rivers sends down from the great forests through which they flow thousands of fallen trees, many of which are very large. This is especially the case with the Yukon River, which empties its immense accumulation of débris into Norton Sound, and the Kuskoquin, emptying into a bay of the same name one hundred and fifty miles farther south. When these tree trunks find their way to the open sea, the prevailing currents bear them southward to the Aleutian Islands, where a large number become stranded at Attoo, and are promptly secured and stored for use as fuel. It would seem to be rather a precarious source of supply to depend upon for this purpose, but we were told that, as a rule, it was ample to meet the demand. There is also a stocky vine growing in great abundance upon the islands, which the native women gather and dry, and this makes a quick, strong fire. At certain seasons the women may be seen in long lines coming from the hills, each one bearing upon her back a monster bundle of this product, which they store for use when the other source of fuel fails them or proves insufficient. The people of Attoo have tamed the wild goose, of which they rear considerable flocks for domestic use, similar to our New England custom with the tame bird, and it is said they are the only tribe in Alaska who do so. Long since the blue fox was by some means introduced upon the island, and being at first properly protected, the place has become fairly stocked with them, a certain number only being killed annually by the natives, and from their valuable fur these Aleuts realize quite a large sum. Were it necessary, lumber could be brought in small quantities from the island of Kodiak, or even from the mainland far away; but there is very little use for it in Attoo, the houses being built of drift-logs and not of boards. Besides the low, thrifty species of shrubbery growing on these islands, there are also wild berries in great abundance, the original seeds having probably been brought by the birds from the mainland. Grasses grow luxuriantly, being cut and cured to feed a few small Siberian cattle through the winter months, though it is hardly necessary to house them at all. They are kept on only one or two of the larger islands of the group. Domestic animals might do well here with a little care, but the attention of the natives is given almost exclusively to the products of the sea, whose very bounty demoralizes them. At Unalaska, of this same group, the natural grass grows to six feet in height, and with such body that one must part it by exerting considerable force in order to get through. The natives braid it into useful and ornamental articles, hats, baskets, mats, and the like. This prolific growth is represented to be remarkably nutritious, and cattle are very fond of it. W. H. Dall predicted that this Aleutian district will yet furnish California with its best butter and cheese; while Dr. Kellogg, botanist of the United States Exploring Expedition, wrote: "Unalaska abounds in grasses, with a climate better adapted for haying than the coast of Oregon. The cattle are remarkably fat, and the milk abundant." This is the refitting station for all vessels passing between the Pacific Ocean and Behring Strait, and here also is the principal trading post of the Alaska Commercial Company.

Mr. George Wardman, United States Treasury Agent, that stated on his late visit to this island he saw in one warehouse sea-otter skins ready for shipment which were worth quarter of a million dollars in the London market. This will represent, perhaps, two thirds of all this class of pelts furnished to the world annually, as comparatively few go from any other quarter. Other land furs are brought here for shipment to San Francisco, two fur companies having headquarters at Unalaska. The place has some sixty native houses, and perhaps five hundred inhabitants. Unalaska is known to be rich in both gold and silver mines, one of which is owned by a San Francisco company, and which it is proposed to fully develop and work during the coming year, careful tests having proven its prospective value.

The same fertility seen at Unalaska exists also at Kodiak and Atagnak, where the small breed of cattle that live upon the grass are as fat as seals, and require no shelter all the year round. There is a small ship-yard near the first named island, where vessels of twenty-five and thirty tons are built for fishing in the neighboring sea. These two islands, situated just off the eastern shore of the Alaska Peninsula, are called the garden spots of this region, enjoying more sunshine and fair weather than any other part of the Territory. They contain rich pastures, beautiful woodlands, and broad open fields, which during the summer are carpeted with constant verdure and wild flowers. Kodiak was for a long time the capital of the Russian American possessions, but the government headquarters were removed for some reason to Sitka. On Wood Island, opposite Kodiak, is the clear and spacious lake which so long furnished ice to the dwellers on the Pacific coast, but particularly to the people of San Francisco. The whole range of Aleutian Islands from Attoo to Kodiak contains between four and five thousand inhabitants, nearly all of whom are called Christians, being members of the Greek Church. They are very generally half-breeds, that is, born of intermarriage between emigrant Russians and native women. Professor Davidson was struck by the strong resemblance of the aboriginal tribes inhabiting these islands to the Chinese and Japanese, and was satisfied that they came originally from Asia. There are many very intelligent persons among them. "They are docile, honest, industrious, and very ingenious," says Professor Davidson. The women of Unalaska have always been noted for the beauty and variety of their woven grass mats and various other ornamental work, particularly in the combinations of colors and unique designs.

This cunning of the hand and artistic ingenuity is not confined to the women; the men are also skillful carvers and engravers. Whenever they have been afforded a fair degree of instruction, and the opportunity to exercise their ability, they have proved themselves to be adepts especially in this last mentioned branch of skilled labor. We have seen artistic work produced by a native Unalaskan which it was difficult to believe was not the performance of some experienced and thoroughly educated European.

The thirty-eight charts in the Hydrographic Atlas of Tebenkoff were all drawn and engraved on copper by a native Aleut.

On the island of Unga, one of the Shumagin group, situated half way between Unalaska and Kodiak, is a small settlement of a score of white men and about a hundred and fifty natives. By a regulation of our Treasury Department, only natives are allowed to hunt the sea-otter, and therefore these white men have married native wives, thereby becoming natives in the eyes of the law. The revenue derived from the sea-otter trade on this island is said to average from six to seven hundred dollars a year to every family. Off the southern shore of the Shumagin group is the best cod fishing bank that is known. It is estimated that a million good-sized cod were taken here last season and shipped to San Francisco. This metropolis of California once depended upon the product of our Newfoundland fisheries for its salted cod, but has drawn its supply for the last few years almost entirely from the coast of Alaska, and the consumption has increased every year.

CHAPTER IX.

Cook's Inlet. — Manufacture of Quass. — Native Piety. — Mummies. — The North Coast. — Geographical Position. — Shallowness of Behring Sea. — Alaskan Peninsula. — Size of Alaska. — A "Terra Incognita." — Reasons why Russia sold it to our Government. — The Price Comparatively Nothing. — Rental of the Seal Islands. — Mr. Seward's Purchase turns out to be a Bonanza.

COOK'S INLET, which lies to the north of the island of Kodiak, was esteemed by the Russians to be the pleasantest portion of Alaska in the summer season, with its bright skies and well wooded shores. It stretches far inland in a northeasterly direction, and is quite out of the region of the fogs which prevail on the coast. Gold has been profitably mined for some years on the Kakny River, which empties into the eastern side of this extensive inlet, and good coal abounds in the neighborhood.

When the Russians first came to this region they taught the natives to make what they called quass, a cooling and comparatively harmless acid drink. To produce this article rye meal is mixed with water, in certain proportions, and allowed to remain in a cask until fermentation takes place and it is sour and lively enough to draw. Latterly the natives have learned to add sugar, and thus to produce a fermented liquor of an intoxicating nature. Progress in this direction has been made until now they mix a certain portion each of sugar, flour, dried apples, and a few hops, when they can be obtained, putting the whole into a close barrel or cask. When fermentation has taken place and the mixture has worked itself clear, it forms a strong intoxicant. This article proves the cause of a thousand ills among the aborigines. In each of the scattered villages among the islands there is sure to be seen a few broken-down victims of this active poison, who have impoverished their families and wrecked their own constitutions.

In each of these Aleutian islands there is found a Russian-Greek chapel and a regularly appointed priest, this religion being preferred by the natives to that of all other sects, captivating their simple minds by its gorgeous show and its mystery. Their honest devotion, however, to a religion which they cannot comprehend may be reasonably questioned. There can be no doubt that their idolatrous customs and original pantheism have been almost entirely abandoned, — ceremonies which were elaborately described by the early voyagers, and which involved strange incantations and even human sacrifices. Intercourse with the whites has at least had the effect of abolishing the most objectionable features of their early superstitions. The bishop of the organization is a Russian and resides in San Francisco, whence he controls these parishes, which he occasionally visits, being amply supplied with pecuniary means by the home government at St. Petersburg. The piety of these Aleuts is very pronounced, so far as all outward observances go, and we were told that they never sit down to their meals without briefly asking a blessing upon their rude repast. Golovin, a Russian who lived many years among the Aleuts, says: "Their attention during religious services is unflinching, though they do not understand a word of the whole rite." The same author goes on to say, "During my ten years' stay in Unalaska not a single case of murder happened among the Aleutians. Not an attempt to kill, nor fight, nor even a considerable dispute, although I often saw them drunk." Hunting is the principal source of their support, and to get the sea-otter they often make long, exposed trips in their undecked boats, and experience many trying hardships. When they return to their homes at the close of the season, having been nearly always reasonably successful, the quass barrel is brought into requisition, and its contents partaken of to excess, drunken orgies following with all their attendant evils.

The Aleuts are a very honest people, quite unlike the Eskimos of the north, who are natural pilferers. They are also possessed of a certain stoicism which compels admiration. When they are sick or suffering great pain they utter no complaint, and outwardly are always content, no matter what the future may send as their lot. An Aleut is never known to sigh, groan, or shed a tear. If he feels it, he never evinces immoderate joy, but is always quiet, moderate, and grave. They are in a great degree fatalists, and believe that which is decreed by the power in the sky will come to pass, whatever they may do to prevent it. It is Kismet.

It is an interesting fact that before these islands were discovered by the Russians, the natives were in the practice of preserving their dead in the form of mummies, and this had probably been their habit for centuries. Satisfactory evidence is afforded by what is found upon the islands to show that they have been the residence of populous tribes for over two thousand years. Mr. Dall, in his indefatigable researches, was able to secure several examples of the mummified dead on these outlying islands, eleven of which came from one cave on the south end of Unalaska, but none were ever found or known to have existed upon the mainland. This fact is looked upon by ethnologists as an important addition to our knowledge of the prehistoric condition of these peculiar people of the far Northwest, now part and parcel of our widespread population. The mummies of Peru and those of Alaska are now arranged side by side in the cases of the Smithsonian Institution at Washington, and what is very singular is that they seem, in their general appearance, to be almost identical.

The interior of Alaska and its more Arctic regions north of the valley of the Yukon remain still only partially explored. No more is actually known of it than of Central Africa. It would be anything but a pleasure excursion, at present, to penetrate the extreme northern harbors of the extended coast line, which are mostly uninhabited, and which are tempest-swept for a large portion of the year. Northwestern Alaska shares with northeastern Siberia the possession of the coldest winter climate in the world, but we must remember it is not always winter, and thousands of Eskimos here find life quite tolerable. Beyond 70° of north latitude no trees are to be found; even shrubs have disappeared, giving place to a scanty growth of lichens and creeping wood-plants. Even here, however, Nature asserts her prerogative and brings forth a few bright flowers and blooming grasses in the brief midsummer days. Point Barrow is what might be termed, in common parlance, "the jumping-off place;" the beginning of that mysterious ocean where the compass needle, which lies horizontal at the equator, attracted by an unexplained influence dips and points straight downward. There is no lack of animal life in this frozen region, the sea is as full as in the tropics; the whale here finds its birthplace, and herring issue forth in countless columns to seek more southern seas, while the air is darkened by innumerable flocks of sea-fowl. The wolves, the polar bear, and other fur-bearing animals afford meat and clothing to the Eskimo to an extent far exceeding his requirements. Only thoroughly organized expeditions and a few adventurous whalers attempt to pass Point Barrow, a long reach of low barren land, and the most northerly portion of the Territory, which projects itself into the great Arctic Ocean very much after the fashion of the North Cape of Norway, in the eastern hemisphere, at latitude 71° 10'.

There is a village at Point Barrow containing about a hundred and fifty people, living in houses partly under ground as a protection against the cold. The roofs are supported by rafters of whale jaws and ribs. This people we call the Eskimo proper. They have a severe climate to contend with, but are abundantly supplied with food and oil from the sea. They have a strange aversion to salt, and any food thus cooked or preserved they will not eat unless driven to it by dire necessity. Our government is just about to erect a comfortable structure here as a sort of refuge to shipwrecked navigators of the Polar Sea, this being the verge of those unknown waters which guard the secret of the Pole.

A peninsula makes out from near the centre of the western coast of Alaska, the terminus of which is the nearest point between this continent and Asia, the two being separated by Behring Strait, where the East and the West confront each other, and where the extreme western boundary of our country is the line which separates Asia from America. This is called Cape Prince of Wales, a rocky point rising in its highest peak to twenty-five hundred feet above the sea. Here is a village of Eskimos numbering between three and four hundred souls, who do not bear a good reputation. They are skilled as fishermen on the sea and hunters on the land, to which it may be added that they are professional smugglers. Here it is quite possible in clear weather to see the Asiatic coast — Eastern Siberia — from United States soil, the distance across the strait being about forty miles. There are two islands in the strait, known as the Diomedes, almost in a direct line between Cape Prince of Wales on one side and East Cape on the other; stepping-stones, as it were, between the two continents. Occasional intercourse between the natives of the two opposite shores is maintained to-day by means of sailing craft, and doubtless has been going on for hundreds, if not for thousands, of years. So moderate are the seas, and so calm the weather hereabouts at some portions of the year, that the passage is made in open or undecked boats.

On King's Island, fifty miles south of Cape Prince of Wales, there is a tribe of veritable cave-dwellers. The island is a great mass of rock, with almost perpendicular sides rising seven hundred feet above the sea. On one side, where the angle is nearly forty-five degrees, the Eskimos have excavated homes in the rock, about half a hundred of which are two hundred feet above the sea. These people openly defy the revenue laws, and are the known distributers of contraband articles, especially of intoxicants.

Behring Sea, where it washes the shores of Alaska, from Norton Sound to Bristol Bay, is slowly growing more shallow, having but fifteen fathoms depth, in some places, forty miles off the west shore of the mainland, and growing shallower as it approaches the continent. This has caused a speculative writer to suggest the possible joining of Asia and America, at some future period, by the gradual filling up of Behring Sea. The reason of this is obvious. The Yukon River brings down from its course of two thousand miles and more many hundred tons of soil daily which it deposits along the coast, while the Kuskoquin River, second only to the Yukon in volume, is engaged in the same work about a hundred and fifty miles south of where the greater river empties into Norton Sound. These large water-ways carry, like the Mississippi, immense deposits to the sea, and the process has been going on night and day for no human being knows how long.

One hundred and fifty miles from the mouth of this Kuskoquin River the Moravians of Bethlehem, Pa., support a missionary establishment. The station is named Bethel, one of the most isolated points in Alaska, receiving a mail but once a year! Truly, nothing save fulfilling a conscientious sense of duty could compensate intelligent people for thus separating themselves from home and friends.

We have spoken of a peninsula making out at the north towards Asia, but this comparatively insignificant projection from the mainland should not be permitted to confuse the reader's mind as regards the Alaska Peninsula, properly so called, which extends from the southern part of the Territory, ending in the islands which form the Aleutian group. This peninsula is undoubtedly one of the most remarkable in the world, being fifty miles broad and three hundred long, literally piled with mountains, some of which are but partially extinct volcanoes, emitting at the present time more or less smoke and ashes, sometimes accompanied by blazing gases discernible at night far away over land and sea, appearing to the midnight watch on board ship like a raging conflagration in the heavens. The principal islands of the group of which we have been speaking, and which stretch far away from the southwestern corner of the Alaska Peninsula towards Kamschatka, as though extending a cordial hand from the Occident to the Orient, are as follows: Unimak, with a volcanic peak nine thousand feet high; Unalaska, whose peak is five thousand seven hundred feet high; Atka, with a height of four thousand eight hundred feet; Kyska, which is crowned by an elevation of three thousand seven hundred feet; and Attoo, whose tallest peak is over three thousand feet. This island is just about four hundred miles from the Asiatic coast. Unimak has a large lake of sulphur within its borders, and all of these islands have more or less hot springs. From those in Unalaska loud reports issue at intervals, like the boom of cannon, recalling our late similar experience in the Yellowstone Park.

Alaska constitutes the northwestern portion of the American continent, and has a coast line exceeding eleven thousand miles. The extreme length of the Territory, north and south, is eleven hundred miles, and its breadth is eight hundred. It is bounded on the north by the Arctic Ocean, on the east by British Columbia, on the south by the Pacific Ocean, and on the west by Behring Strait and the North Pacific. Our geographies and encyclopædias help us to little more than the boundaries of this great Territory, which contains nearly six hundred thousand square miles. The latest published estimates give the aggregate number of square miles as nineteen thousand less than the amount we have named, but Governor Swineford and other residents of the Territory believe it to be an underestimate. As there is no actual survey extant, the figures given can only be a reasonable approximation to the true number. The boundary dividing Alaska and British Columbia was settled by treaty between England and Russia in 1825, and the same line is recognized to-day as separating our possessions in this quarter from those of Great Britain. Alaska is as large as all of the New England and Middle States, with Ohio, Indiana, Illinois, Wisconsin, Michigan, Kentucky, and Tennessee combined. So far as size is concerned, the Territory is, therefore, an empire in itself, being equal in area to seventy-one States like Massachusetts, and containing as many square miles as England, Ireland, Scotland, Wales, France, Spain, Portugal, Switzerland, and Belgium united. It has been estimated by competent judges that, with its islands, it has a coast line equal to the circumference of the globe. Very few of our people, even among the educated class, have an adequate idea of the immensity of this northwestern Territory, two thirds of which abounds in available resources, only awaiting development. Were Alaska situated on our Atlantic coast it would extend from Maine to Florida.

Miss Kate Field, in a comprehensive article already quoted from, published in the "North American Review," justly censuring Congress for its supineness and ignorance in relation to Alaska, says: "American citizens, living comfortably on the Atlantic seaboard, knowing their own wants and dictating terms to their submissive representatives, take little heed of those new additions to the United States which are destined to be the crowning glory of the Republic. When a nation is so big as to render portions of it a *terra incognita* to those who make the laws, there's something rotten this side of Denmark! . . . The march of empire goes on in spite of human fallibility, and now the land of the midnight sun knocks at the door of Congress. She is twenty-three years old, and asks to be treated as though she were of age. The big-wigs at Washington rub their eyes, put on their spectacles, and wonder what this Hyperborean hubbub means?"

In examining the geographical characteristics of Alaska, we observe a peculiarity in its outlying islands which is also found in the construction of the continents. They all have east of their southern points series of islands. Thus, Alaska has the Sitkan or Alexander group; Africa has Madagascar; Asia has Ceylon; Australia has the two large islands of New Zealand; and America has the Falkland Islands. Alaska is the great island region of the United States.

It is not for us to enter into the brief history of the country, that is, brief as known to us, but it is well to fix in the mind the fact that Russia's title was derived from prior discovery. Behring first saw the continent in this region of North America, July 18, 1741, in latitude 58° 28', and two days later anchored in a bay near a point which he called St. Elias, a name which he also gave to the great mountain overshadowing the neighboring shore. It is sufficient for our purpose that we know this Territory was purchased from Russia by our government in 1867, after that country had occupied it a little more than a century, paying therefor the sum of seven million two hundred thousand dollars. It has been truly said that it was practically giving away the country on the part of Russia; but doubtless diplomatic reasons influenced the Tzar, who would much rather have presented it outright to the United States than to have it, by conquest or otherwise, fall into the hands of England, who was known to crave its possession as connected with her Pacific coast interests. So when the first Napoleon sold us Louisana, he did so not alone in consideration of the money, which was doubtless much needed by his treasury,—amounting to sixty million francs,— but because he was not willing to leave this distant territory a prey to Great Britain in the event of hostilities between France and England, which were then imminent. He was glad, as he remarked, "to establish forever the power of the United States, and give to England a maritime rival destined to humble her pride;" adding, "It is for the interest of France that America should be great and strong."

Alaska was a white elephant to Russia, but in our hands it has already proved a bonanza.

Any one can now see that the sum named as an equivalent for this colossal territory was a trifling value to place upon it, when its great extent is realized, together with its vast mineral wealth and inexhaustible supply of fish, fur, and timber. It is in fact the only great game and fur preserve left in the Western world, inviting the trapper and hunter to reap a rich return for their industry. Nowhere else on this continent do wild animals more abound, or enjoy such immunity from harm, as is afforded them in the dense, half-impenetrable forests of Alaska, where Nature herself becomes our gamekeeper, preventing the too rapid extinction of animal life.

From a lease in favor of the Alaska Commercial Company of San Francisco, giving them the exclusive right to take seals on the Prybiloff group of islands, our government has received four and one half per cent. interest, annually, during the last nineteen years, on the entire purchase-money paid to Russia. This same company, whose term is just about to expire, would gladly renew the lease with our government at a considerable advance upon the amount heretofore paid; but it is an open question whether the continuance of this great monopoly is for the best interest of Alaska, when considered in all its bearings.

Undoubtedly this contract is a real benefit in one way. The company, through its agents, will take good care to see that no outside interest interferes with their rights so as to permit any indiscriminate slaughter of the seals. Whereas, were the capture of these peltries not guarded, an end of the product would be brought about in a very short time. There is a manifest injustice in all monopolies, as we view them; but of two evils, in this instance we should perhaps feel inclined to choose the least by selling the privilege to a responsible company. It must be admitted that the high-handed course of the present company, their arbitrary assumptions, and their treatment of the natives generally, are represented in a very bad light by many residents of Alaska; but little else, however, could be expected of so great a monopoly. One thing is certain, and that is, the company has realized a great fortune by its contract.

There were plenty of people who ridiculed the acquisition of this Territory at the time when it was brought about; but there were also some far-seeing statesmen, influenced by no selfish motives, who felt very different about the matter, among whom was Mr. Seward, then Secretary of State, and to whom the credit is mostly due for consummating the important purchase. That able diplomat considered the transaction to have been the most important act of his official career, and put himself on record to that effect. He remarked, in discussing the matter at a public meeting, " It may take two generations before the purchase is properly appreciated." Mr. Seward was right. It was a crowning glory for him to have added a new empire to his country's domain, though in 1867 its great commercial importance was hardly known, even to himself. Its valuable gold deposits were then thought possibly to exist; but subsequent developments have already far outstripped anticipations in that direction, and the large yield of the precious metal is annually increasing.

" I thought when Alaska was purchased, in 1867," says that keen observer and clever writer, Captain John Codman, " that it might answer for a great skating park; but now I know, from merely coasting along its southeastern shores and landing at a few of its outposts, that the seven million two hundred thousand dollars paid for it is less than the interest of the sum that it is worth. A great part of it is yet unexplored, for its whole area is three times greater than the republic of France; but what has been discovered is invaluable, and what has not been discovered may be valuable beyond calculation."

So little did we, as a people, appreciate the new acquisition that it was almost entirely neglected for seventeen years. Not until 1884 was it granted a territorial government, Hon. John H. Kinkead, ex-governor of Nevada, being the first governor appointed for Alaska. "Twenty years ago," says Governor Swineford of Alaska, "I made political capital out of Seward's purchase. I called it the refrigerator of the United States. I heaped obloquy on William H. Seward. I shall spend the rest of my life in making reparation to what I have so foully wronged." Such has been the general testimony of all who speak from personal observation, and uninfluenced by sinister motives.

CHAPTER X.

Territorial Acquisitions. — Population of Alaska. — Steady Commercial Growth. — Primeval Forests. — The Country teems with Animal Life. — A Mighty Reserve of Codfish. — Native Food. — Fur-Bearing Animals. — Islands of St. George and St. Paul. — Interesting Habits of the Fur-Seal. — The Breeding Season. — Their Natural Food. — Mammoth Size of the Bull Seals.

THE subject of the addition of Alaska to the United States suggests the fact that our territorial acquisitions from time to time form certain decided and interesting landmarks in the history of the country. Thus, in 1803 we acquired Louisiana from France by the payment of fifteen million dollars. In 1845 Texas was annexed and her debt assumed, amounting to the sum of seven million five hundred thousand dollars. In 1848 California, New Mexico, and Utah were acquired from Mexico, partly through war, and by the payment of fifteen million dollars. In 1854 Arizona was purchased from Mexico for ten million dollars. And last, but by no means least, Alaska, as has been stated, was obtained from Russia in 1867 for seven million two hundred thousand dollars. "By this purchase," said Charles Sumner in his able speech before Congress, "we dismiss one more monarch from this continent. One by one they have retired; first France; then Spain;

then France again ; and now Russia ; all give way to the absorbing Unity which is declared in the national motto, *E Pluribus Unum.*"

At the time of the transfer of Alaska, the native population, Russians, half-breeds and all, did not probably exceed forty thousand ; indeed, careful inquiry seems to indicate that this is an overestimate. Since that period the native population has steadily decreased, but the white population has increased, it is believed, sufficiently to make good the estimated aggregate of twenty-two years ago. In 1867 the commerce of Alaska was officially reported as being two million five hundred thousand dollars for the current year. The published estimate for the last year made it a fraction less than seven million dollars, of which about a million five hundred thousand dollars was in gold bullion. Certainly this shows a very steady if not rapid commercial growth. Competent individuals estimate that the commerce of the Territory for the year 1889 will reach ten million dollars in amount. The increase in the number of fish-canning establishments alone will add two millions to last year's aggregate. The shipment of preserved salmon exported in tins and barrels is increasing annually.

The available timber now standing in the Territory might alone meet the ordinary demand of this continent for half a century. Though the extreme northern part of Alaska is treeless, its southern shores, both of the islands and mainland, are covered with a dense forest growth, the Aleutian group excepted. It is the visible wealth of the country, and a source of admiration to all appreciative visitors.

Fort Tongas is very near the southeast point of Alaska, and about ten miles north of Fort Simpson; the former American, the latter English territory. When the ground was cleared to establish the American fort, "yellow cedar-trees," says W. H. Dall, "eight feet in diameter were cut down. The flanks of all the islands of this archipelago bear a magnificent growth of the finest timber, from the water's edge to fifteen hundred feet above the sea." It must be a cedar of magnificent proportions out of which the natives can hew and construct a canoe seventy feet long capable of carrying one hundred men. This the Haidas do, producing models both swift and seaworthy, the prows extending in a peak not unlike the ancient galleys of Greece, decorated with totemic designs. These magnificent forests, having never felt the stroke of the axe, present a growth naturally very dense and peculiar, the branches of the tall trees being often draped with long black and white moss, dry and fine as hair, which it resembles. This characteristic recalled the same effect observed upon the thickly wooded shores of the St. John River in Florida, and the Lake Pontchartrain district of Louisiana. The fallen trees and stumps are cushioned by a growth of green, velvety moss, nearly ten inches in thickness, and are also decked with creeping vines in the most picturesque manner; among which is seen here and there deep red clusters of the bunch-berry. The timber is pronounced by good judges to be as valuable as that of Oregon and Washington, compared with which our forests in Maine are hardly more than tall undergrowth. A very large percentage of the Alaska timber grows at the most convenient points for shipment, making it especially available. The white spruce, called the Sitka pine, rises to a height of from a hundred and fifty to a hundred and eighty feet, and measures from three to six feet in diameter. When this growth is cut into dimension lumber it very much resembles our southern pitch-pine. . There is also found in these forests the usual variety of cedar, fir, ash, maple, and birch trees, mingled with the others of loftier growth. The yellow cedar of this region grows nowhere else of such size and quality. It is much prized, and best adapted for shipbuilding, having been found to be unequaled for durability, and also because it is impervious to the troublesome teredo, or boring worm, which destroys the ordinary piles under the wharves at Puget Sound, as well as at Sitka, so rapidly as to render it necessary to renew them every three or four years. Southern latitudes, in the neighborhood of the Gulf of Mexico, suffer equally from the depredations of this active marine pest. The Alaska cedar is also a choice cabinet wood, possessing a very agreeable odor, considerable quantities of it being shipped for select use in San Francisco and elsewhere. The coast of the Alexander Archipelago comprises nearly eight thousand miles of shore line, forming long straight avenues of calm deep water many miles in length, sprinkled with islands densely wooded from the water's edge, while the number of good harbors is almost countless, in which vessels may lay alongside the land and receive their cargoes of timber or lumber in the most convenient manner.

When the woods of Maine and Michigan cease to yield satisfactorily, as they must do by and by, we have here a ready source of supply which no ordinary demand can exhaust in many years. One enthusiastic writer upon this subject predicts that this part of the North Pacific coast will eventually become the ship-yard of the American continent. One is hardly prepared to indorse so sweeping a prediction, but that there is a nearly inexhaustible supply of the necessary timber for such a purpose even an inexperienced visitor cannot fail to realize. It is gratifying to know that these forests are free from all danger by fire, which often proves so destructive in the State of Washington and elsewhere. This immunity from a much dreaded exigency is owing to the frequent rains, which keep the undergrowth in Alaska so moist that the flames cannot spread.

Speaking of Fort Tongas, we should not forget to mention that a native couple, educated by the missionaries, are here teaching a school of young natives numbering fifty pupils, for which our government pays them five hundred dollars per annum. The success attained by these instructors in teaching the ordinary branches of an English education is surprising. Tongas, it will be remembered, is the most southerly point of our Alaska possessions.

The country teems with animal life. The sea which laves its shores and the outlying islands is so full of excellent fish as to have been a wonder in this respect since the days of the earliest navigators. The same may be said of its rivers, inlets, and lakes, the former being famous for the abundance, size, and excellence of the salmon which they produce, and which are annually packed for exportation in such large quantities to various parts of the world. We were told by the overseer of the canning factory at Pyramid Harbor that the entire product of the establishment was already — the season but just commencing — engaged by a Liverpool house. To secure the delivery the foreign merchant had cheerfully advanced five hundred pounds sterling.

"The Alaska banks would be an ocean paradise to the Newfoundland fishermen," says Professor Davidson. "The eastern part of Behring Sea 'is a mighty reserve of cod,' and the area within the limits of fifty fathoms of water is no less than eighteen thousand miles." "What I have seen," said W. H. Seward at Sitka, in 1869, "has almost made me a convert to the theory of some naturalists, that the waters of the globe are filled with stores for the sustenance of animal life surpassing the available productions of the land." The coast also abounds in oysters, clams, mussels, and crabs. The oysters are small, but of excellent flavor, and might be greatly improved by cultivation. Clams and mussels are much esteemed by the aborigines, the first-named being large and of prime quality. They dry the clams, as they do salmon and cod, using no salt in the process, but stringing them by the score on long blades of strong grass, and in this shape laying them away for winter use. There is certainly some special preservative quality in the atmosphere here which enables the natives to keep clams unfrozen in good condition for several months. The matter of "ripeness," however, makes no difference to these Indians, who seem actually to prefer their fish a little putrid, and oil is purposely kept until it becomes so before they will use it.

The hills and valleys of the islands and the mainland support more fur-bearing animals than can be found on any other part of this continent, and we certainly believe of any other part of the world. The great variety includes bears of several species, wolves, beavers, deer, foxes, caribou, martens, mountain goats, moose, musk-oxen, and others. Herds of walruses are found on the far north coast, as well as in Behring Sea, which yield food to the natives, and the best of ivory for sale to the traders. It is a curious fact that no reptile, toad, lizard, or similar animal is to be found in Alaskan territory. The waters of the North Pacific, from the most westerly of the Aleutian Islands up to Behring Strait, swarm with cod, haddock, sturgeon, large flounders, and halibut, while our hardy whalemen successfully pursue their mammoth game both north and south of the strait. When the country was first discovered, there was another important animal found here in considerable numbers, known as the sea-cow, which furnished Vancouver and his crew with wholesome and palatable meat, and which had formed a source of food supply for the aborigines probably for centuries. But this large, amphibious animal, thirty feet long and seal-like in shape, has now entirely disappeared. This was owing to merciless slaughter by the Russians, who found the sea-cow an easy prey to capture, because of its inactivity and clumsiness in the water, besides which, the creature is said to have been utterly fearless of man, making no effort to escape when attacked. They are represented to have been fierce when attacked by the wolves, and to have been fully able to defend themselves.

Two islands lying to the north of the Aleutian group form a favorite resort of the fur-seal, which so abounds in this region that nearly a century of active war waged upon them by the hunters, for the sake of their valuable skins, has produced no perceptible diminution in their numbers. This is partly owing, however, to the fact that of late years the killing has been restricted as to the aggregate annual number, and also as to the sex and age of the seals. The pelts sent from Alaska have not fallen short of a hundred thousand annually for the last twenty years, and it is believed by those who should be able to judge correctly that this number has been very much exceeded.

There is hardly an uninterested person in the Territory who will not express this opinion.

The two islands referred to in Behring Sea, namely, St. Paul and St. George, together with two smaller and unimportant ones named respectively Otter Island, which is situated six miles south of St. Paul, and Walrus Island, about the same distance to the eastward, are known as the Prybiloff group. St. Paul is thirteen miles long by four broad; St. George is ten miles long and between four and five broad. Neither of them have any harbor in which vessels can safely lie, but they anchor half a mile or more off shore, and freight is taken or delivered by means of lighters. So violent is the surf at times on these islands in mid-ocean that if the wind is unfavorable no attempt at landing is made. Otter Island is peculiar in being nothing more nor less than an extinct volcano, with a still gaping, threatening crater, and an elevation of three hundred feet above the surrounding sea. Its only occupants consist of water-fowl and blue foxes, both as plentiful as peas in a pod. The animals were introduced long ago for breeding purposes, and have greatly increased. These are the "seal islands" so often spoken of, and which furnish four fifths of all the sealskins used in the markets of the world. This sounds like an extravagant estimate, but it is believed to be quite correct.

The islands are of volcanic origin, having been thrown up from the bottom of the sea in comparatively modern times. When one speaks of geological facts, one or two thousand years are considered very brief periods. At the time of their discovery, St. George and St. Paul were uninhabited, but native Aleuts, the nearest of whom lived about two hundred miles south of these islands, were brought hither and domesticated, to work for the Russian Fur Company. Since the transfer to our government these people have worked uninterruptedly for the Alaska Commercial Company, which has, in addition to the headquarters of the seal-fishery, some forty trading stations in the Territory.

We speak of the "seal-fisheries," but there is in reality no fishing about the business. The seals are all taken on land. The employees of the company get between the seals and the water and drive such as are selected inland like a flock of sheep. They move slowly, pulling themselves along by their fore flippers, as a dog might do with his hind legs broken, but they get over the ground at the rate of one or two miles in the hour, and are driven the latter distance to the warehouse before the killing takes place.

It is curious that these two islands only, with a few small spots in the North Pacific, should possess the peculiar conditions of landing-ground and climate combined which are necessary for the perfect life and reproduction of the fur-seal. H. W. Elliott, who acted as United States government agent for four seasons at the seal islands, and who is good authority upon this special subject, says: "With the exception of these seal islands of Behring Sea, there are none elsewhere in the world of the slightest importance to-day. When, therefore, we note the eagerness with which our civilization calls for sealskin fur, in spite of fashion and its caprices, and the fact that it is and always will be an article of intrinsic value and in demand, it at once occurs to us that the government is exceedingly fortunate in having this great amphibious stock-yard, far up and away in this seclusion of Behring Sea, from which it can draw continuous revenue, and on which its wise regulations and its firm hand can continue the seals forever."

This writer's remarks should be qualified, however, so far as to state that the Russians possess some profitable " rookeries " situated on the Commander Islands, seven hundred miles to the southwest of the Prybiloff group, where the same policy of protection for breeding purposes is enforced as govern the traffic on our own islands. It is true that the product of the Russian islands is as nothing compared with that of St. Paul and St. George. A small number of fur-seal are also secured on the coast of Brazil, and at the Shetland and Falkland Islands, giving perhaps twenty thousand pelts annually from other sources than those named in Alaska. It is our own opinion that at least forty thousand pelts are sent to market by unauthorized people from the islands and coast of Alaska, which number should be added to the hundred thousand which the regular company are entitled to export, in getting at the aggregate produced by the Territory.

The two seal islands leased to the Alaska Commercial Company are about thirty miles apart, and are seemingly among the most insignificant landmarks known in the ocean. It is only on very modern maps that they are designated at all, but they afford to the seals the happiest isolation and shelter, their position being such as to envelop them in fog banks nine days out of ten during the entire season of resort. Neither the seals nor the natives can long bear the glare of the summer sun, and so find no fault with this prevailing screen between them and the sky. There are no icebergs, properly so called, in these waters. Behring Strait is too shallow for anything but light field ice to pass into the North Pacific or Behring Sea; there is therefore no fear of visits from the polar bears often seen floating about in the frozen sea at the north. They would make sad havoc among the seals were they to get so far south, and drive them away altogether. Ice floats off from the immediate shores in the spring, but encountering the thermal current, this soon dissolves, and is no impediment to navigation. It is marvelous that the natives dwelling on the group do not die of the poisoned atmosphere arising from the thousands upon thousands of seal carcasses annually slaughtered, and which are left to decay upon the ground. The stench thus created is so powerful that vessels sailing to leeward, three or four miles off shore, are permeated by it, and though their captains may not have been able to get a solar observation for many days, they can easily tell their exact latitude and longitude by "dead reckoning." Naval surgeons have been detached by government to visit and examine the physical condition of the people on St. George and St. Paul, touching this very matter, and they have reported that the natives enjoyed good health, the mortality among them being at a very low average compared with that of other semi-civilized communities favorably situated. There is a church and school-house on each of the islands, with white teachers, and also a skilled physician, who is paid for his services by the Commercial Company.

The fur-seal traffic has heretofore exceeded all other regular business in value conducted in this Territory, though the product of the precious metals will in future probably take the lead, hard pressed by the rapidly growing development of the fisheries. The habits of the seal are interesting and very peculiar. It is a social animal, and evinces a degree of intelligence nearly approaching that of the dog. Occasionally a young one is found domesticated among the natives of the more populous islands, and when thus brought up among human beings they become very tractable, and are easily taught many amusing tricks. They move in herds, coming to the breeding grounds in large numbers, and at regular periods of the year, that is in the latter part of May and early in June. The contrast between the male and female seal is great, the former being large, bold, and aggressive, the latter small, peaceful, and quiet; both are models of grace and symmetry after their kind. While the males are specimens of great physical strength, the females are delicate, timid, and affectionate. The young are born blind and so remain for a couple of weeks, or more. When they are about six weeks old the mother takes them into the water to teach them to swim. They are very shy of the sea at first, but persistent effort on the mother's part soon makes them expert swimmers, and rapidly develops that side of their nature. During the breeding season the old males remain on shore, fasting all the while, and growing extremely thin, living by absorption of the blubber which they accumulate while at sea, so that upon retiring at the end of the season they are but a mere shadow of their former selves. They return again the next season, however, as plethoric as ever.

"All the bulls," says Mr. Elliott, "from the very first, that have been able to hold their positions, have not left them from the moment of their landing, for a single instant, night or day; nor will they do so until the end of the rutting season, which subsides entirely between August 1st and 10th. It begins shortly after the coming of the cows in early June. Of necessity, therefore, this causes them to fast, to abstain entirely from food of any kind, or water, for three months at least; and a few of them actually stay out four months, in total abstinence, before going back into the ocean for the first time after 'hauling up.' They then return as so many bony shadows of what they were a few months previously, covered with wounds; abject and spiritless, they laboriously crawl back to the sea to obtain a fresh lease of life."

The natural food of the seal is believed to be small fishes and kelp, that prolific product of the ocean which is found floating in nearly all latitudes, being torn from its rocky bed by storms and carried everywhere on the tides and currents. The females seldom give birth to more than one at a time, and though they are naturally a very docile animal, the mother will fight savagely for her young. The old males weigh from two to three hundred pounds each, when they first land, soon gathering a harem about them of a dozen females or more, and permitting no other bull to approach the circle. There are occasional elopements among the females, enticed away by young bachelor seals, who have no family ties to occupy them, but as a rule the females remain loyal, at least during the season. The full grown male reaches seven feet in length, and the female about five feet; the latter averages about a hundred pounds in weight, the former weigh twice as much and often more. Nature seems to produce a much larger number of females than of males, besides which the law protects the female from the hunter. The killing of these animals on St. Paul and St. George is nearly all done in six weeks of each year, say from the 10th of June to the 20th of July. As regards the fur, a seal at four years of age is thought to yield the best, and is therefore considered to be at that time in his prime. It is the males of this age, accordingly, which are selected for slaughter. So numerous are these animals that the shore is often black with them, three or four thousand being in sight within the space of a hundred square rods. The pups are full of playfulness, rolling and tumbling about like a litter of kittens. The rule not to kill the old bulls and female young is a necessary precaution to prevent the extermination of the race, which indiscriminate slaughter has probably done in so many other places.

CHAPTER XI.

Enormous Slaughter of Seals. — Manner of Killing. — Battles between the Bulls. — A Mythical Island. — The Seal as Food. — The Sea-Otter. — A Rare and Valuable Fur. — The Baby Sea-Otter. — Great Breeding-Place of Birds. — Banks of the Yukon River. — Fur-Bearing Land Animals. — Aggregate Value of the Trade. — Character of the Native Race.

SURGEON J. B. PARKER tells us in a published article upon the fur-seals of Alaska, that just previous to the transfer of the country to this government five hundred thousand sealskins were being taken from these islands annually, though it was pretended by the Russians that they restricted the number to one quarter of this total. The strange instinct of the animals which causes them to return yearly in such marvelous numbers to be slaughtered is a mystery difficult to solve. Persistent cruelty exercised towards them for a century has not disturbed their affection for this chosen breeding-place of their ancestors in Behring Sea.

The seals are universally killed by a sharp blow upon the head from a club, which fractures the skull and produces instant death. The natives are so skillful in dealing this blow that a second one is not necessary, and the seal cannot reasonably be supposed to suffer any pain, so that the operation is robbed of all cruel features. The frequent battles fought between the old bulls to maintain possession of their chosen ground and their harems are represented to be of the fiercest character, sometimes ending in the death of one of the combatants, though they are so very hardy and tenacious of life that this is by no means common. The breeding season is at its height in the middle of July. Early in September, the pups having learned to swim, the "rookeries" are gradually broken up for the season, old and young departing together for the deep-sea feeding grounds, nothing being seen of them again as a body until the following May or June. It is quite a mystery as to where they go, but that they promptly disperse in various directions seems most probable, as no seals are met with in large numbers by navigators of the Pacific or the South Seas, and they only land for breeding purposes. The author has seen a few in the month of March off the Samoan group of islands, also in the month of December near the coast of Cochin China. And again, in crossing the Indian Ocean from Bombay to the mouth of the Red Sea, in February, an occasional head of the fur-seal would appear above the surface of the ocean, showing how widely dispersed these animals are. There is a theory which has long existed, to the effect that when the seals depart from Behring Sea they seek a lonely island group in the central Pacific Ocean, somewhere between 53° and 55° north latitude, and longitude 160° to 170° west, where they pass their winter months in peace and plenty. Expeditions have been fitted out at

San Francisco for the purpose of discovering these possible islands, but no one has ever seen them. Those most conversant with seal-life do not entertain this supposition, and for good reasons. If any such land existed in the region designated it would surely have been discovered, as it is too near the direct track of commerce not to have been sighted long ago.

The flesh of the fur-seal is eaten by the natives, and the blubber also serves for fuel, as well as furnishing a much-used oil. The stench of the burning fat is extremely disgusting to one not accustomed to it. There is but little lean meat on the animal; nearly the whole body is composed of blubber. The whites eat the flesh of the young seal, which is not unpalatable when properly prepared, and is called Alaska pork. When the females arrive at the "rookeries," like the old males, they are in remarkably good flesh, so much so, indeed, as to render locomotion difficult; but though they do not fast like the bulls, they nevertheless become quite thin by the end of the season.

St. George and St. Paul islands contain about three hundred and fifty Aleuts, whose sole business is killing and skinning the seals, and afterwards salting and packing the pelts for shipment. They are all in the regular employment of the Commercial Company, which leases the islands. By the terms of the lease from our government, only natives of the Aleutian group of islands can be employed to kill the seals; no whites except the overseers are permitted to remain on the two islands.

An agent of the United States occasionally visits them to see that the spirit of the lease is faithfully adhered to; otherwise they are quite isolated from the outer world. Under the protective system, which is presumedly adhered to, the number of seals is said to be on the increase, and the space on the shores which they occupy is enlarged yearly. It has been officially estimated, after actual inspection, that over one million seals are born on these islands every year. It is asserted that double the number of pelts now authorized could safely be taken from the Pribyloff group annually, and it would certainly seem so, when this extraordinary fecundity is realized. But it must also be taken into consideration, that man is not the only enemy which the fur-seal has to encounter. When the young ones leave the shore to begin their deep-sea life, they become the prey of many marine cormorants, among which the shark is said to be the most active. This tiger of the ocean does not attack the large, full-grown seals, who are too wary and active for him, but the young ones often fill his capacious maw.

The aborigines employed upon the seal islands do not reach a very old age; persons of over fifty years are seldom found among them. Consumption is the most fatal disease which they encounter; this runs its course with singular speed after being once contracted. All attempts of the physicians are in vain; the patient, falling into a condition of hopeless indifference, soon passes away. We were told that the natives of Alaska generally were very difficult to treat medically, ignoring the benefit of medicines, and generally refusing to take them. These semi-savages will not hesitate to resort to incantations to exorcise evil spirits (or disease, which to them is the same thing), but they fear to use the white man's agent to remove these evil influences.

For a number of years the manufacture of oil from seal blubber was followed by the fur company with profit, thus disposing of the carcasses of the animals whose skin had been removed; but oil-making on the seal islands has been discontinued, as being no longer a paying business.

The sea-otter is a large animal, having fine, close black fur, sprinkled with long, white-tipped hairs, which strongly individualize it and add much to its beauty. Its pelt is used mostly for trimming, being both too heavy and too expensive for making up into entire garments. The size of a full-grown skin is about four feet in length by about two and a half wide. It is a solitary marine animal, never seen in numbers, rarely even with a mate, and is extremely shy, demanding great patience and shrewdness in the hunter to insure its capture. This animal rarely lands except to bring forth its young, and the natives say that it sometimes gives birth to its progeny on floating sedge or kelp at sea. Of this material the ingenious creature makes a sort of buoyant nest, according to the natives' ideas. When sleeping, it floats upon its back, carrying its young clasped to its body in a ludicrously human fashion. The

Indians hunt the animals by going out a considerable distance to sea in their frail canoes, and watching for the appearance of the otter's nose above the water, they paddle silently towards it so as not to disturb the game. At the proper moment the well-balanced and delicate lance is thrown with unerring aim. A careful watch is then kept for the reappearance of the otter, which must soon come to the surface to breathe, being a warm-blooded, respiratory animal. A second lance is pretty sure to disable the otter, when it floats helpless on the surface, falling an easy prey to the pursuer. At times six or eight natives in single canoes join in the hunt, so as to form a broad circle; the nearest one to the otter when he rises after being wounded is the one to throw the second lance. The hunters obtain from the local traders between forty and fifty dollars for a full-grown otter skin, and sometimes double that amount, so that if successful in the pursuit they are well rewarded for many hours of patient watchfulness, aside from which they realize a keen enjoyment in the pursuit as sportsmen.

The hunters oftenest pursue their game alone, and if a native secures an otter after a whole week of watching he feels well repaid, though during that time he has lived on a scanty supply of food, and has slept nightly in the open air exposed to the rain. Sometimes his watch is kept in his boat upon the sea, and sometimes among the rocks on the shore, in a bay where the otters are known to resort occasionally. A few years of such rough life and exposure ages even an Alaskan Indian, and it is not surprising that rheumatism and consumption should so prevail among them. Up to a certain stage such a life may harden the hunter, but the turning-point comes at last, and when the native begins to fail in physical strength he does so rapidly; simply giving way to the first attack, rejecting all medicine which the white man may offer, and unless he is an important member of his tribe, a chief or a leader of some sort, even the shaman or medicine man with his incantations is not called in. Good nursing is discarded, the invalid considers it to be his fate to die, and seems to go half way to meet the grim destroyer.

The fur of the sea-otter varies in beauty of texture and value according to the animal's age and the season of the year in which it is captured. They are considered to be in their prime when about five years old, and those skins which are taken in winter are always of a more beautiful texture than those which are secured in summer. Of all animals hunted by man it is most on the alert, and, as we have said, most difficult to obtain. One intelligent statement declares that before they were so systematically hunted eight thousand skins were shipped from Alaska in a single year, but we believe that from four to five thousand otter skins would be considered a good twelve months' yield in these days. The Saanack islets and reefs are the principal resort of these animals on the coast, and hither the natives come from long distances to hunt them, camping on the main island. Frequent attempts have been made to rear the young sea-otter, specimens being often taken when the mother is captured, but they always perish by starvation, never partaking of food after being separated from the mother; a well-known fact, which was referred to with not a little sentiment by the experienced hunter who related the circumstance to us. "Him die of broke heart," said the native, attempting an expression of tenderness upon his egg-shaped features, which proved a ludicrous caricature. We saw a stuffed specimen of a young sea-otter in a native cabin at Juneau, consisting of the skin only, but very cleverly mounted and preserved by the hunter who had captured its mother.

It is somewhat singular that the world's supply of otter fur, like that of sealskin, comes almost entirely from the coast of Alaska, in the North Pacific and Behring Sea. Otter fur may be said to be almost confined in its geographical distribution to the northwest shores of America.

The successful pursuit of the animal, so far as the natives are concerned, is of even more importance than that of the fur-seals, for contingent upon its chase, and from the proceeds of its pelts, some five thousand natives are enabled to live in comparative luxury. It requires, as we have shown, great energy, hardihood, and patient application to effect its capture, but the sea-otter is a most beneficent gift of Providence to these aborigines, and administers, as well, to the pride of the fashionable world. The natives in former times attached great importance to preparing themselves for hunting the sea-otter, fasting, bathing, and performing certain mystic rites before embarking for the purpose. After his return from a successful hunt the Aleut was accustomed to destroy the garments which he wore during the expedition, throwing them into the sea, so that the otters might find them and come to the conclusion that their late persecutor had been drowned and there was no further danger in frequenting the shore. This practice, ridiculous as it seems to us, serves to illustrate the superstitious character of the Alaskan natives, who seldom fail to see omens in the most trifling every-day occurrences.

The interior and northern parts of Alaska are the greatest breeding-places for birds in the world, being the resort of innumerable flocks, which come from various parts of this continent, and others which make the tropical islands their home a large portion of the year on both the Atlantic and Pacific sides of America. These myriads of the feathered tribes consist largely of geese, ducks, and swans, coming hither for nesting, and to fatten upon the wild salmon berries, red and black currants, cranberries, blackberries, bilberries, and the like, which greatly abound during the brief but intense Arctic summer. There are eleven kinds of edible berries which mature in August, among which the wild strawberries are the finest flavored we have ever eaten. It is said that the geese especially become so fat feeding upon the plentiful supply of wholesome food that at the close of the season they can hardly fly, and are thus easily caught by the natives, who, in turn, feast luxuriously upon their tender and succulent flesh. Explorers tell us that they have seen on the banks of the Yukon — the great river of central Alaska, and the third in magnitude in America — the breeding-place of the canvasback ducks, which has been heretofore a matter of some mystery. They prepare on the banks of this northern watercourse broad platforms of sedge, mingled with small twigs and bushes, laid compactly on marshy places, and without building a carefully arranged nest deposit their eggs in untold numbers. That keen and scientific observer, the late Major Kennicott, says he saw on the banks of the Yukon acres of marshy ground thus covered with the eggs of the canvasback ducks, in numbers defying computation. "The region drained by the Upper Yukon is spoken of by explorers," says Mr. Charles Hallock, editor of "Forest and Stream," "as being a perfect Eden, where flowers bloom, beneficent plants yield their berries and fruits, majestic trees spread their umbrageous fronds, and songbirds make the branches vocal. The water of the streams is pure and pellucid; the blue of the rippled lake is like Geneva's; their banks resplendent with verdure, and with grass and shining pebbles."

At the first approach of winter the augmented

millions of birds take flight for the low latitudes, or their homes in the temperate zone, the old birds accompanied by the broods which they have hatched in the solitudes of the far north. Those which have come from the neighborhood of the Caribbean Sea turn in their flight unerringly in that direction; those from the South Pacific islands heading as surely for that tropical region. Only the ptarmigan and the Arctic owl, with a few of the white-hawk family, remain to brave the winter cold of northern Alaska, with the hardy Eskimo, the walrus, and the polar bear. The smaller tribes of birds are well represented here in the summer season, even including several species of swallows, martins, and sparrows, these tiny creatures seeming to follow some general bird instinct. Even the domestic robin is seen as far north as Sitka. Limited scientific research has recognized and classified one hundred and ninety-two different kinds of birds which are found in this Territory, a considerable number of which were unknown to science previous to 1867.

We have said nothing relative to the hair-seals, or sea-lions, of Alaska, because their importance is comparatively insignificant, having no commercial value. Nevertheless, they are utilized by the ingenious natives in various ways; the hides serve as a covering for a certain class of boats, made with wooden frames, and are also employed for several domestic purposes. The walrus is found in largest numbers on the north coast, in the true Arctic region, affording some valuable oil, together with considerable ivory, in carving which the natives are very expert. Though the fur-trade of the land is by no means equal to that of the sea, still its aggregate results are very considerable. It employs numerous hunters and gives profitable business to many white traders, nearly all of whom make a permanent home in the Territory. Undoubtedly the most prolific and valuable fur-yielding district on the mainland is the valley of the Yukon, where the beaver, marten, several kinds of bears, with the wolf and fox, afford the best fur. We saw at the principal store in Wrangel many packages of bearskins prepared for shipment to San Francisco. These packages would average five hundred dollars each in value, and had been gathered from those brought in by the natives during the two weeks intervening between the arrival of the regular steamers. Single bearskins sell here, according to their marketable character, for from twenty-five to thirty-five dollars each. The natives make little or no use of these skins, preferring the woolen blanket of commerce. The red and cross fox is found everywhere in the Territory, and its skin is comparatively cheap. It is singular that the blue fox is found only on the islands of St. Paul, St. George, Attoo, and Atkha, while the white fox is to be sought only at the far north. There is also the black fox, which, however, is a great rarity, thought to be an occasional accident of nature; the skins always bring extravagant prices from the traders. The black fox is not found in any special locality, but occurs now and again in any part of the Territory. The skin of the silver fox is also highly prized, and proves a valuable peltry to the native hunters, forty dollars each being the usual price paid by the white traders. Only a few hundred are taken yearly. The land-otter and the beaver so abound as to make up a large total value annually. The latest official records show that there has been produced and shipped from Alaska annually an average of fifty-seven thousand beaver skins; eighteen thousand land-otter skins; seventy-one thousand foxes' skins of the various sorts; and of musk-rats two hundred and twenty-one thousand. These figures should be largely added to in each instance (we were told by one official that this aggregate estimate should be doubled), in order to include the unregistered pelts which are annually secured by various hunters, both whites and natives, and which find their way to distant markets through irregular channels, more especially over the borders of British Columbia.

This fur-trade is open to all, but requires capital, organization, and persistency to make it profitable. The natives do nearly all of the hunting and trapping, and will only engage in it, as a rule, to supply themselves with means to procure certain luxuries from the trader's store, such as sugar, tea, and tobacco. We are sorry to add to these comparative necessities the article of whiskey, which is only too often furnished illicitly to the eager natives. When these wants are supplied they idle away their time until stimulated once more by their necessities to go upon the trail of the fur-bearing animals. Of course there are some exceptions to this, many of them being steady and willing workers, but we speak of the average native. There is no fear of the supply of furs being exhausted under this system of capture; even a combined and vigorous effort on the part of the hunters could not accomplish that in many years. Unlike our western Indians, these Alaskans are a comparatively thrifty race, entirely self-sustaining, and never require support from the government, notwithstanding idleness is their besetting sin, as is, indeed, characteristic of uncivilized people everywhere.

We were told of several of these aborigines who had learned the lesson of thrift from the whites to such good effect as to have saved sums of money varying from one to five hundred dollars, which they had deposited in the Savings Bank of San Francisco, and upon which they drew their annual interest; an investment, the safety and economy of which they fully appreciated.

CHAPTER XII.

Climate of Alaska. — Ample Grass for Domestic Cattle. — Winter and Summer Seasons. — The Japanese Current. — Temperature in the Interior. — The Eskimos. — Their Customs. — Their Homes. — These Arctic Regions once Tropical. — The Mississippi of Alaska. — Placer Mines. — The Natives. — Strong Inclination for Intoxicants.

IT is a well-known fact, proven by official observations, that the climate of the Pacific coast is considerably more temperate than that of the same latitude on the Atlantic side of the continent. The record of ten consecutive years, kept at Sitka, gave an annual mean of 46° Fah.

This is in latitude 57° 3′ north, and is found by comparison to be four degrees warmer than the average of Portland, Me., or six degrees warmer than the temperature of Quebec, Canada. The average winter is milder, therefore, at Sitka than it is at Boston, however singular the assertion may at first strike us, in connection with the commonly entertained idea of this northwestern Territory. The mean winter temperature of Sitka and Newport, R. I., are very nearly the same, and there is only a difference of six degrees in their mean yearly temperature, though there is a difference of sixteen degrees of latitude.

We have before us a printed letter which appeared in the "Philadelphia Press," signed by

Mr. C. F. Fowler, late an agent of the Alaska Fur Company, who has resided for twelve years in Alaska, in which he says: "You who live in the States look upon this country as a land of perpetual ice and snow, yet I grew in my garden last year, at Kodiak, abundant crops of radishes, lettuce, carrots, onions, cauliflowers, cabbages, peas, turnips, potatoes, beets, parsnips, and celery. Within five miles of this garden was one of the largest glaciers in Alaska." In a certain sense it is surely a country of paradoxes.

The harbor of Sitka is never closed by ice, which cannot be truthfully said of Boston or New York.

Dr. Sheldon Jackson, long resident in the Territory as United States general agent of education for Alaska, tells us that the temperature of Sitka and that of Richmond, Va., are nearly identical. Mr. McLean of the United States Signal Service, who has been located at Sitka for several years, says, "the climate of southern Alaska is the most equable I ever experienced."

There is in Alaska a very large section of country, composed of islands and the mainland, where the average temperature is higher than at Christiania, capital of Norway, or Stockholm, capital of Sweden, — where the winters are milder and the fall of rain and snow is less than in southern Scandinavia, which is the geographical counterpart of Alaska in the opposite hemisphere. Sitka harbor is no more subject to arctic temperature than is Chesapeake Bay. "It must be a fastidious person," said Mr. Seward in his speech upon Alaska, " who complains of a climate in which, while the eagle delights to soar, the hummingbird does not disdain to flutter." If it is sometimes misty and foggy on the coast, it is not so to a greater extent than is the case during a large portion of the year in the cities of London and Liverpool.

Both the islands and mainland of this latitude afford ample grass for cows, sheep, and horses, also producing, with ordinary care, the usual domestic vegetables, as we have shown, the assertion of certain writers to the contrary notwithstanding. We have not far to look for the cause of this favorable temperature existing at so northerly a range of latitude. The thermal stream known as the Japanese Current, coming from the far south charged with equatorial heat, is precisely similar in its effect to that of the better known Gulf Stream on our Atlantic coast, rendering the climate of these islands and the coast of the mainland of the North Pacific remarkably warm and humid. We speak especially and at length of this subject of the temperature of Alaska, because a wrong impression is so generally held concerning it. At a distance from the coast the temperature falls, and most of the inland rivers are closed by ice half the year. Even in the interior we are in about the same latitude and average temperature of St. Petersburg. Thus on the line of Behring Strait the annual mean at Fort Yukon, which lies just inside of the Arctic circle, six hundred miles inland from Norton Sound, is 16.92°; this is in latitude 64° north. Along the coast of southern Alaska the fall of snow is not greater in amount than is experienced during an ordinary winter in the New England States, and it disappears even more quickly than it does in Vermont and New Hampshire. In the interior and at the far north, the quantity of snow is of course much greater, and covers the ground for about half the year.

But where the sun shines continuously throughout the twenty-four hours, the growth of vegetable life is extremely rapid. The snow has hardly disappeared before a mass of herbage springs up, and on the spot so lately covered by a white sheet, sparkling with frosty crystals, there is spread a soft mantle of variegated green. The leaves, blossoms, and fruits rapidly follow each other, so that even in this boreal region there is seed-time and harvest. The annual recurrence of this carnival season is all the more impressive in the realm of the Frost King.

The Japanese Current, already referred to, strikes these shores at Queen Charlotte Island in latitude 50° north, where it divides, one portion going northward and westward along the coast of Alaska, and the other southward, tempering the waters which border upon Washington, Oregon, and California; hence their mild climate. Sea captains who frequently make the voyage between San Francisco and Yokohama have told the author that this Japanese Current — with banks and bottom of cold water, while its body and surface are warm — is so clearly defined as to be distinguishable in color from the ordinary hue of the Pacific Ocean, and that its deep blue forms a visible line of demarcation between the greater body and itself along its entire course. The thermometer will easily define such a current, and this the author has often seen demonstrated from a ship's deck; but it must be a very keen eye that can distinguish such differences of color at sea as the above assertion would indicate.

In so extended a territory as that of Alaska, with broad plains, deep valleys, and lofty mountain ranges, it is reasonable to suppose there must be a great diversity of climate. The brief inland summer is represented to exhibit marked extremes of heat, and the winter corresponding extremes of cold. W. H. Dall, an undoubted authority in all matters relating to the valley of the Yukon, though his book upon the country was published some twenty years since, says: "At Fort Yukon I have seen the thermometer at noon, not in the direct rays of the sun, stand at 112°, and I was informed by the commander of the post that several spirit thermometers graded up to 120° had burst under the scorching sun of the Arctic midsummer." Fort Yukon is the most northerly point in Alaska inhabited by white men. It is estimated that ten or twelve thousand Eskimos live in the uninviting region north of the Yukon valley. They are a most remarkable people, who are struggling with the cold three quarters of the year, and who seem to be strangely content with a bare existence. Their days and nights, their seasons and years, are not like those of the rest of the world. Six months of day is succeeded by six months of night. They have three months of sunless winter, three months of nightless summer, and six months of gloomy twilight. No Christian enlightenment or religious teaching of any sort has ever found its way into this region. The people believe in evil spirits and powers who are in some way to be propitiated, but have no conception of a Divine Being who overrules all things for good. Like the southern Alaskans they are superstitious to the last degree, and discover omens in the most ordinary occurrences. The decencies of life are almost totally disregarded among them, their highest purpose being apparently the achievement of animal comfort and gorging themselves with food and oil.

Their sky is famous for its beautiful auroral display — gorgeous pyrotechnics of nature — in the long, chill winter night, when a brilliant arch spans the heavens from east to west, marked with oscillating hues of yellow, blue, green, and violet, rendering everything light as day for a few moments, then falling back into darkness. So off the coast of Norway among the Lofoden Islands, the hardy men who pursue the cod-fishery in that region, during the winter season, depend upon the Aurora Borealis to light their midnight labor, that being considered the most favorable hour of the twenty-four to secure the fish. Without this nocturnal meteoric illumination, it would be darkness indeed in the polar regions for half the year.

This phenomenon in its Arctic development is so much intensified as to quite belittle the exhibition with which we are familiar in New England, and which is called the Northern Lights.

It is certainly very odd that these boreal natives, the Eskimos proper, should have precisely the same mode of salutation which the New Zealand Maoris practice, though they are separated by so many thousand miles of ocean, namely, the rubbing of noses together between two persons who desire to evince pleasure at meeting. No matter how oily the Eskimo's nose may be, or however dirty the Maori's face, to decline this mode of salutation when offered is to give mortal offense, either in tropical New Zealand or in arctic Alaska, at Point Barrow or at Ohinemutu. "The home of the Eskimos," says Bancroft, in his excellent work on the natives of the Pacific coast, "is a model of filth and freeness. Coyness is not one of their vices, nor is modesty ranked among their virtues. The latitude of innocency characterizes all their social relations; they refuse to do nothing in public that they would do in private." They seem to live in a primitive state, without craving anything of the white man's possessions, except tobacco and rum, which are smuggled to them by contrabandists, who come on to the coast to trade for furs and ivory. This class of traders, sailing from San Francisco, and stopping at the Hawaiian Islands to procure a few hogsheads of the vilest intoxicant which is made, pass along the northern coast of Alaska, touching at certain places where they are expected annually. The walrus not only supplies the Eskimo with food, but its tusks are used as the common currency among them, and are secured in considerable quantities by the illicit traders. The encroachment of unscrupulous contrabandists renders the utter extinction of the walrus only a question of time. It is to be regretted that the wholesale slaughter of this animal cannot be prevented. If this could be brought about, as in the instance of the fur-seal, we might continue to get ivory from the shores of the Frozen Sea for all time. The natural enemy of the walrus is the polar bear, but his most relentless pursuer is man.

These Eskimos wrap their dead in skins closely sewed and lay them in the tundra, together with the worldly possessions of the deceased, without any funeral ceremonies. It would be sacrilege for any one to disturb this property left with the body, and no member of the tribe would think of doing so.

In the Yukon Valley the remains of elephants and buffaloes are found fossilized, as those of the rhinoceros were discovered on the opposite continent in Siberia, thus showing that this now arctic region was once tropical, a conclusion, nevertheless, which seems to be almost impossible to the traveler while gazing upon Niagaras of frozen rivers in the month of July.

The Yukon River is the Mississippi of Alaska, forming with its several tributaries the great inland highway of the Territory. As yet there are no roads in the country, everything is transported by water or on the backs of the natives; the great importance of such an extensive water-way can therefore be readily understood. The magnitude of the Yukon — one of the twelve longest rivers in the world — will be realized by the fact that it is still a matter of doubt among different writers which of the two rivers named is the largest with respect to the volume of their currents, though Ivan Petroff, in his report as agent of the Secretary of the Interior, speaks thus confidently upon the subject: " The people of the United States will not be quick to take the idea that the volume of water in an Alaskan river is greater than that discharged by their own Mississippi; but it is entirely within the bounds of honest statement to say that the Yukon River, the vast deltoid mouth of which opens into Norton Sound, of Behring Strait, discharges every hour of recorded time as much, if not one third more, water, than the 'Father of Waters' as it flows to the Gulf of Mexico."

This writer does not seem to us given to exaggeration, but still we are a little inclined to question the accuracy of his estimate as to the volume of water borne seaward by this great Alaskan river.

The Yukon rises in the Rocky Mountain range of British Columbia; entering Alaska at about 64° north latitude, and pursuing its course nearly from east to west across the entire Territory, it finally empties, as stated, into Behring Strait through Norton Sound. The river is navigable for fifteen to eighteen hundred miles, while its entire length is computed at over two thousand miles, with an average width of five miles for half the distance from its mouth. There are several places on the lower Yukon where one bank is invisible from the other. It is seventy-five miles across its five mouths and the intersecting deltas. At some places, six or seven hundred miles inland, the river expands to twenty miles in breadth, thus forming in the interior a series of connected lakes, which explorers pronounce to be deep and navigable in all parts. This great water-way can only be said to have been partially explored, but those persevering pioneers who have made the attempt to unravel its mysteries have given us extremely interesting details of their experiences, all uniting in bearing witness that its banks are rich in fur-bearing animals, and that its waters are stocked with an abundance of fish, including the all-pervading salmon. These valuable fishes follow the same instinct which they exhibit in other parts of the world, in their annual pilgrimage of reproduction, that is, after entering a river's mouth, to advance as far as possible towards its source. Besides fish and fur-bearing animals, the region through which the Yukon flows contains abundant deposits of gold, silver, copper, nickel, and bituminous coal. Some placer gold mines which were worked on its banks and in its shallows, so long as the season permitted, are credibly reported to have yielded to one party of prospectors nearly eighty dollars per day to each man.

The trouble to be encountered in working these placers is owing to their remoteness from all sources of supply, and the exposure to the long winters which prevail in the placer gold-producing regions. These are obstacles, however, which will one of these days be overcome by the erection of suitable shelter, and a rich new mining field will thus be permanently opened. There are a number of trading-posts along the course of the Yukon at which white men reside permanently to traffic with the natives, purchasing furs from such as will hunt; and there are many who are represented to be industrious and provident, supplying the whites with meat and fish as well as with pelts, fully appreciating the advantage of steady habits and regular wages. In this respect the inland tribes differ materially from most of those living on the coast; the latter care little for work or wages until they are driven by necessity to seek employment. We speak in general terms; there are of course many worthy exceptions, but savage races have little idea of thrift, and like the wild animals are aroused to action only by the demands of hunger. In equatorial regions where the nutritious fruits are so abundant that the natives have only to pluck and to eat, they are sluggish, dirty, and heedless, living only for the present hour. In this Arctic region where the sea is crowded with food and the fields are covered with berries, the same listlessness prevails as regards the future with nine out of ten of the aborigines. These remarks do not apply to the Aleuts, from whom the Commercial Company obtains its workmen. These are mostly half-breeds, who are far more civilized than are our Western Indians.

The proprietors of the Treadwell gold mine, Douglass Island, and of the works at Silver Bow Basin, employ large numbers of the natives, finding them to be reliable and industrious laborers.

"Where we can separate these Alaskan natives from the objectionable influences which are apt to grow up in populous centres, and especially from multitudes of adventurous miners who come from a distance, we find them to be faithful and tractable workers," said an employer to us.

"How about the Chinese?" we asked.

"They are excellent workers," was the reply. "Set them a task, show them how to perform it, and it will surely be done. They are almost like automatons in this respect and require no watching."

"Then why not employ them more generally?"

"Because of the prejudice, the unreasonable prejudice, against them. Our other workmen rebel if we keep many Chinamen on the pay-roll."

This corresponded exactly with the author's experience elsewhere, in various parts of the world where the Chinese have sought a new home outside of China. John is not perfect, but he is infinitely superior to a large portion of the drinking, rowdy, and restless foreign element which fills so large a place in the labor field of this country.

The greatest care is necessary to keep spirituous liquors away from the aborigines, a craving for which is beyond their control where there is a possibility of its being obtained. When they fall under its influence they seem to utterly lose their senses, and become dangerous both to themselves and to the whites. As has been intimated, the only means of locomotion is afforded by the watercourses, and the natives, being excellent canoeists, find ample employment of this nature, both in traversing the rivers and along the shore of the islands. The waters of the Yukon, like those of the Neva at St. Petersburg, freeze to a depth of five or six feet in winter.

CHAPTER XIII.

Sailing Northward. — Chinese Labor. — Unexplored Islands. — The Alexander Archipelago. — Rich Virgin Soil. — Fish Canning. — Myriads of Salmon. — Native Villages. — Reckless Habits. — Awkward Fashions and their Origin. — Tattooing Young Girls. — Peculiar Effect of Inland Passages. — Mountain Echoes. — Moonlight and Midnight on the Sea.

LET us observe more order in these notes, and resume the course of our experiences in consecutive form.

As we speed on our sinuous course northward, inhaling with delight the pure and balmy atmosphere, bearing always a little westerly, winding through narrow channels which divide the richly wooded wilderness of islands, avoiding here and there the ambuscaded reefs, the pleasurable sensation is intense. The scenery, while in some respects similar to that of the St. Lawrence River and the Hudson of New York, is yet infinitely superior to either. After having reached latitude 54° 40' we come upon Dixon Entrance, a reach of the sea which separates Alaska from British Columbia, and from this point we are sailing exclusively in the purple shadow of our own shore, and in the waters of the United States. At times we pass islands as large as the State of Massachusetts, whose picturesque and irregular mountainous surfaces are covered with immemorial trees, and whose unknown interiors are believed to be rich in coal, iron, silver, and other metals. The axe has never echoed in the deep shade of these dense plantations of nature; they form a pathless wilderness, solemn and silent, save for the stealthy tread of wild beasts, the mournful music of waving pines, and the occasional notes of wandering seabirds. The migratory flocks of the tropics as a rule go farther north to raise their broods, but a few, weary of wing, shorten their aerial journey and build nests on these islands. For many centuries past the great columnar trees have grown to mammoth size, and have then fallen only by the weight of years, enriching the ground with their decayed substance and giving place to another similar growth, which, in its turn, has also flourished and passed away. How like the course of human races! This process has been going on perhaps for twice ten thousand years. "Nature alone is antique," says Carlyle. The past history of Alaska, except for a comparatively short period, is a blank to the people of the nineteenth century.

Day after day there is a continuous and unbroken chain of mountain scenery. On the right of our course is a broad strip of the mainland, an Alpine region, thirty miles in width, which forms a part of southern Alaska, bounded on the east by British Columbia, and on the west by the many spacious islands, which create so perfect a breakwater that the constant swell of the contiguous ocean is not felt. Some of these islands lie within a quarter of a mile of each other, on either side of our way, and yet the water is far too deep to admit of anchoring, the peaks rising abruptly from unknown depths to thousands of feet above the sea. The channels seem still more narrow from the great height of the mountains which line the course. The eye catches with delight the bright ribbons of waterfalls tumbling down their sides, in gleeful uproar, foaming and sparkling towards the depths below. These are fed by melting snow and hidden lakes far up in the cloud-screened summits. Some of these waterfalls, narrow and swift, leap from point to point, now forming small cascades, and now continuing in a perpendicular form like a column of crystal. Others, so abrupt and precipitous are the heights from which they are launched, fall in an unbroken stream, clinging to the cliffs at first, but quickly expanding into a thin sheet rivaling the Bridal Veil of the Yosemite, and reaching the base in a constant gauzelike spray.

The wide, open tracks seen now and then on the steep, thickly-wooded mountain sides, reaching from high up to the snow-line down to the very surface of the water, are the pathways swept by giant avalanches. What immense power and lightning-like speed are suggested by the broad, clean swath that is left! The wind caused by the rushing avalanches is almost equally resistless, the trees on either side of the track being torn into splinters by it.

Now and again, above the tops of the giant

pines, one can see moving objects on the exposed peaks and cliffs, almost too far away and too small for identification, but we know them to be wild mountain goats, — the Alaskan chamois, — quite safe from the hunters in these perilous heights, never trod by the foot of man. The tender glow of twilight enshrouding mountain peaks, emerald isles, and the gently throbbing bosom of the sea, added daily a witching charm to a scene which already seemed perfect in beauty.

The principal island group lying off the shore of southwestern Alaska is named the Alexander Archipelago, in honor of the Tzar of Russia. It extends about three hundred miles north and south, and is seventy-five miles from east to west, embracing over eleven hundred islands, scarcely one of which has been explored. The group reaches from Dixon Entrance to Cross Sound, in latitude 58° 25' north. Upon landing at one of these islands it was found to be covered by an impervious forest; the mass of timber and undergrowth was so compact as to defy our progress. The tangle of bushes, roots, vines, and branches formed almost as impenetrable a wall as though built of masonry. The wildest jungles of India are not more dense. Where not covered and hidden by trees, the earth was flecked here and there by the sun, being carpeted with moss and ferns so thickly spread as to form a spongy surface, upon which only the velvety feet of small wild animals could be sustained. A human pedestrian, were he to attempt to pass over it, would sink in this vegetable compound knee-deep at every step. There are no paths in these jungles; the natives have no occasion to penetrate them, their living comes from the sea, and the river courses are their hunting grounds.

This virgin soil, were it to be drained and cleared of trees, would be rich beyond calculation, while the climate is such as to warrant the growth and ripening of any vegetation which will thrive on the Atlantic coast north of Chesapeake Bay. One who has not seen it in Alaska knows not what rank and luxuriant forest undergrowth is. No tropical islands can surpass the Alexander Archipelago in this respect. Thus far no one has come to this region with the idea of testing its availability for agricultural purposes; it is other business which has attracted them. Nothing of any account has ever been done in the way of stock-raising, though the winters of southern Alaska, of Kodiak, and the Aleutian Islands are much milder than are those of Wyoming or northern Dakota, and there is plenty of food for innumerable herds all the year round. If government will but give the Territory of Alaska proper land laws, this region will promptly invite emigration, and be rapidly peopled by thrifty stock-growers.

As we increase our northern latitude forests of tall cedars, spruce, and hemlock still line the shore of the mainland, and cover the countless islands with a mantle of softest green. It is not surprising that artists become enthusiastic over the infinite variety of shades found in these verdant woods, an effect which we have never seen excelled even in equatorial regions. Gliding over the still, deep, pellucid surface of the ocean, we behold these cliffs, forests, and mountains, with coronets of snow reflected therein, as though there was another world below, like that above the rose-tinted sea. One finds almost exactly repeated here the bold, towering peaks, and low-lying rocky isles of the Lofoden group in the far North Sea of the opposite hemisphere, whose sharp, jagged pinnacles have been aptly compared to shark's teeth.

Near Cape Fox, on the mainland, there are two large fish-canning establishments, where salmon are packed in one pound tin cases for shipment to distant markets, and in which a few Chinamen are employed. Some Indian women also find occupation in the establishment, while their husbands capture and bring in the fish in large quantities. This is a rapidly growing and profitable business in this region, there being already forty or fifty such factories along the coast and among the islands north of Cape Fox.

Kasa-an Bay makes into Prince of Wales Island twenty miles, more or less, from Clarence Strait. Here there are several villages of Kasa-an Indians. No spot on the coast is more famous for the abundance and excellence of its salmon; at certain seasons the waters of the bay swarm with them. Here is a large cannery, or fish-packing station, where native women do most of the indoor work. Two thousand barrels of salted salmon were shipped from this bay last year. This was independent of those used in canning. There would seem to be no limit to the expansion of an industry that can furnish such desirable, every way wholesome, and nutritious food to be sold in all parts of the world.

The North Pacific Trading and Packing Company of San Francisco has been doing a profitable business on the coast for many years. In spite of government neglect, commerce is steadily increasing and developing Alaska; it invades all zones, proving the greatest of civilizing agencies. Not only is it the equalizer of the wealth, but also of the intelligence, of nations, and this one branch alone is gradually populating whole districts. When the active packing season is over there is still profitable employment for all. Some are occupied in making the tin cans to hold one pound each; others are taught to become coopers, furnishing the casks for shipping such fish as are split, salted, and exported in that form; while others are occupied in making pine-wood boxes to contain two dozen each of the filled cans. Thus a well-conducted fish-packing establishment employs many people, and presents a busy scene all the year round.

The salmon are so plenty in the regular season that an Indian will sometimes deliver at the canning factory three or four canoe-loads in a single day. They are mostly caught by net or seine, but often during the height of the season the natives absolutely shovel the salmon out of the water and on to the shore with their paddle blades. We were told that as many as three thousand salmon, and even more, are sometimes taken at a single haul of the seine; also that fish of this species weighing from twenty to thirty pounds were common here. Great numbers are discarded at the factories because they do not prove to be of the high pink color which is required by the purchasers and consumers. It seems that the bears know very well when the run of salmon commences, and that there are certain quiet inlets where the fish are sure to get crowded and jammed, so that Bruin has only to reach out his paws and draw one after another on to the shore and eat until he has his fill. The bear-paths leading to these spots are strongly marked, and the animals are thus easily tracked and shot by the hunters. It is the white men who capture them most generally, as the natives have some mysterious reverence and fear combined regarding this animal. They do hunt them, however, but shrive themselves of all sense of wrong by going through some mystic rites. Mr. Charles Hallock says: "There are bears enough in Alaska, grizzly, cinnamon, and black, to furnish every man on the Pacific with a cap and overcoat, and leave breeding stock enough for next year's supply." The grizzly bear is a dangerous animal to encounter single-handed. A bullet seems to have no more effect upon him, unless it strikes a vital spot, than it does upon an elephant. It is necessary to use guns of large calibre when hunting the animal, and the whites rarely seek them unless several tried men band together for the purpose.

From time to time small native villages are seen on the islands and the mainland, all typical of the people, and quite picturesque in their dirtiness and peculiar construction. Some of their cabins are built of boards, but mostly they are rude, bark-covered logs. In front of these dwellings stand totem-poles, presenting hideous faces carved upon them in bold relief, together with uncouth figures of birds, beasts, and fishes. A portion of these tall posts are weather-beaten and neglected, significantly tottering on their foundations, green with mould, unconsciously foreshadowing the fate of the aboriginal race. Groups of natives in bright-colored blankets, with scarlet and yellow handkerchiefs on their heads, come into view, watching us curiously as we glide over the smooth water, while bevies of half-naked children are seen shifting hither and thither in clamorous excitement. What wonderfully bright, black eyes these children have! Some of the women are gathering kelp, for the shores are lined with edible algæ, possessing not only fine nutritious qualities, but being also a recognized tonic, with excellent medicinal properties. This sea-product is collected in the most favorable season of the year, and after being pressed into convenient sized and esculent cakes is stored for future use. The native hamlets are always built near to the shore, accessibility to the water being the first consideration, because from that source comes nine tenths of their subsistence. To clear the forest and secure open fields presupposes more thrift and application than these natives possess; but it would unveil some of the richest soil in the world. These Alaskans have no idea of sewerage, or the proper disposal of domestic refuse. All accumulations of this sort are thrown just outside the doors of their dwellings, to the right and left, anywhere in fact which is handiest. The stench which surrounds their cabins, under these circumstances, is almost unbearable by civilized people, and must be very unwholesome. These natives have broad faces, small, pig-like eyes, and high cheek bones, not very nice to look upon, yet not without a certain expression of real intelligence gleaming through the accumulated dirt.

"What is needed here," said a humorous observer to us, "is the mission teacher with his Bible, spelling-book, and — soap!"

The women cut their hair short on the forehead, nearly even with the eyebrows, causing one to surmise that these Thlinkits — a generic name given to the tribes in this vicinity — must have set the fashion of " banging " the hair, which is so popular among civilized belles. Just so the Japanese women originated the hideous fashion of the "bustle." The author saw this awkward and unbecoming appendage worn upon the backs of the women of Yokohama, Tokio, and Nagasaki three years before it appeared upon the streets of Boston and New York. And now we hear of the " clinging " style of drapery, in which underskirts even are discarded, called the Grecian or classic style. Alas! will nothing but extremes satisfy the importunate demands of fashion? Heaven send that we do not import another fashion from Alaska or the South Seas, namely tattooing. It is quite common here, among young girls of about twelve years of age, whose cheeks and chins are often thus disfigured by irregular lines. The more the natives associate with the whites, however, the more rarely this tattooing is resorted to, and it may be said, as a fashion, to be going out in Alaska, though it is undoubtedly one of the most widely diffused practices of savage life, from the Arctic to the Antarctic circle.

The Alaskans have an original way of producing this indelible marking, the color being fixed by drawing a thread under the skin, whereas the usual mode among various savages is by pricking it in with a needle. The favorite colors are red and blue. We were told that common women were permitted to adorn their chins with but one vertical line in the centre, and one parallel to it on either side, while a woman of the better or wealthier class is allowed two vertical lines from each corner of the mouth. The New Zealand Maori women tattoo their chins in a very similar manner, keeping the rest of the face in a natural condition.

We had threaded the intricate labyrinth of islands, bays, and channels, guarded by miles upon miles of sentinel peaks, nearly all day, on one occasion, under a depressing fog and rain, when suddenly a bold headland was rounded, which had seemed for hours to completely bar our way, and we passed out from under the shadow of the frowning cliffs and the gloom of the dark fathomless waters just as the sun burst forth, warm, bright, and resistless, while the view expanded before us nearly to the horizon. The mist, like shrouded ghosts, stole silently away, vanishing behind the rocks and cliffs. Every dewy drop of moisture, on ship and shore. glittered like diamonds in the dazzling rays of the new-born light, changing the verdant islands into a glory of color, and the whole view to one of majestic loveliness, through which we glided as smoothly as though in a gondola upon the Grand Canal at Venice.

When approaching a landing or anchorage, a signal gun is fired from the forecastle of the ship, creating a series of echoes deep, sonorous, and startling, but especially remarkable for the number of times the sound is repeated. One single gun becomes multiplied to a whole broadside. The report is taken up again and again by other localities, and thus is conveyed for miles away, finally sinking to a whisper, as it were, among the foot-hills of the giant elevations.

The most impressive scenes realized by the traveler are those of moonlight and midnight. How a love of the stars and the sea grows upon one, and life has so few moments of perfect contentment! What melody and magic permeate the pure, placid atmosphere, bounded by the sapphire sea and the azure sky! How tender and beautiful is the utter stillness of the hour! Such scenes of gladness make the heart almost afraid, — afraid lest there should be some keen sorrow lurking in ambush to awaken us from pleasant dreams to the stern, disenchanting experiences of real life.

CHAPTER XIV.

The Alaskan's Habit of Gambling. — Extraordinary Domestic Carvings. — Silver Bracelets. — Prevailing Superstitions. — Disposal of the Dead. — The Native "Potlatch." — Cannibalism. — Ambitions of Preferment. — Human Sacrifices. — The Tribes slowly decreasing in Numbers. — Influence of the Women. — Witchcraft. — Fetich Worship. — The Native Canoes. — Eskimo Skin Boats.

THE aborigines of Alaska are slow in their movements, and in this respect resemble the Lapps of Scandinavia, having also a drawling manner of speech entirely in consonance with their bodily movements. They are as inveterate gamblers as the Chinese, often passing whole days and nights absorbed in the occupation, the result of which is in no way contingent upon intelligence or skill, until finally one of the party walks off winner of all the stakes. Their principal gambling game is played with a handful of small sticks of different colors, which are called by various names, such as the crab, the whale, the duck, and so on. The player shuffles all the sticks together, then counting out a certain number he places them under cover of bunches of moss. The object seems to be to guess in which pile is the whale, and in which the crab, or the duck. Individuals often lose at this seemingly trifling game all their worldly possessions. We were told of instances where, spurred on by excitement, a native risks his wife and children, and if he loses, they become the recognized property of the winner, nor would any one think of interfering with such a settlement. These extreme cases, of course, are rare.

It is impossible to see the aborigines eagerly absorbed in the game without recalling Dr. Johnson's characteristic definition of gambling, namely, "A mode of transferring property without producing any intermediate good."

Inside of the rude native houses one finds many hideous carvings, representing impossible animals and strange objects of all sorts, after the style of the totem-poles, of which we shall have occasion to speak. Many of their small domestic utensils are made from the horns of the mountain goats, and are also curiously carved with nightmare objects, as evil to look upon as African idols. Yet some of these articles show considerable skill and infinite patience in execution. We have seen specimens that it was difficult to believe were executed by the hand of an uncultured savage. Before the Russians introduced iron and steel knives, the aborigines seem to have carved only with copper and stone implements, producing remarkable results under the circumstances. The young women wear silver bracelets, pounded out of American dollar pieces, some of which are an inch broad, and are covered elaborately after civilized models, others bear native heraldic devices of birds, beasts, and fishes, which are said to represent the arms of the wearer's family, it being customary for each tribe and person to adopt some distinctive seal or crest. They much prefer silver ornaments to those of gold or other material; though they are not slow to realize intrinsic values, probably they choose the less expensive metal because it is Alaska fashion.

In spite of all the missionary effort which is made to enlighten these natives, they are still slaves to the most debasing superstitions. Scarcely a month passes in which the civil authorities are not called upon to interfere with the people for cruelty. We were told of one instance which lately occurred at Juneau. A native was seriously ill, and the medicine-man, having failed to relieve him by his noisy incantations, charged an old member of the tribe with having bewitched the invalid. He was consequently seized, tied up, and whipped until nearly insensible, being left for three days without food. By chance the authorities heard of the case and released the old man. The two principal natives who had been guilty of the maltreatment were tried and fined twenty dollars each. The very next day the old man was missing, and it was found that he had again been tied up and whipped. The two culprits admitted repeating their cruelty, saying they had paid for the right to whip out the witch from the old man, and it must be done before the invalid would recover. These ignorant creatures entertained no malice towards the old native; it was only a matter of duty, as they thought, to exorcise the evil one which had possessed the invalid. This is a fair sample of the superstition of the average Alaskans.

When a member of the family dies, the body is not removed for final disposal by the door which the living are accustomed to use, but a plank is torn from the side or back of the dwelling, through which the corpse is passed, after which the place is at once carefully made whole. This, they say, is to prevent the spirit of the defunct from finding its way back again, and thus bringing ill luck upon the living. A still more superstitious and savage custom prevails among some of these ignorant natives.

If a person dies in a cabin, it is held that the place becomes sacred to his spirit, and therefore is unfit for the living. To avoid this difficulty the dying are passed out of the domicile through some temporary hole into the open air to breathe their last, so that neither the house nor the threshold may be sacrificed to the spirit of the dead. Slaves, besides poor widows and orphans, when they die, are often disposed of in the most summary and unfeeling manner, being exposed in the woods, or cast into the sea as food for the fishes. In this connection we remember that the highly civilized and rich Parsees of Bombay do not hesitate to give the dead bodies of their cherished ones to the vultures, in those terrible Towers of Silence on Malabar Hill.

The ceremonies which follow all funerals among these aborigines are peculiar affairs, and for the carrying out of which each person saves more or less of his worldly effects to leave after death. As soon as the body of the deceased is disposed of, then commences what is here called a "potlatch," signifying a "big feast," conducted very much after the style of the New Zealanders on a similar occasion. Everybody is invited and a free spread or feast provided, the same being kept up for several days and nights, so long, indeed, as the purchasing power lasts. Whiskey is freely dispensed, when it can be had, but if not obtainable, as it is a contraband article, then "hoochenoo," made from flour and molasses well fermented, takes its place, being equally intoxicating and maddening. Dancing, wailing, singing, fighting, and grave indecencies follow each other, until the means to keep up the potlatch left by the deceased are exhausted, and his surviving family oftentimes impoverished.

Cremation is the Thlinkit's favorite mode of disposing of his dead. The bodies of slaves and "witches" are disposed of with great secrecy. They are not considered worth burial, and are sometimes cast into the sea, but water burial is infrequent. The bodies of chiefs lie in state several days; the people observe certain rites; then the body is cremated and the ashes are encased in the base of a totem erected to his memory. Shamans (doctors) are never cremated. After lying in state four days, one day in each corner of the cabin, the body is taken out of the house through the smokestack, or some opening other than the door, and conveyed some distance to a deadhouse built for this particular occupant. There in its last resting-place the body is seated in an upright position. The paraphernalia of his rank and office, some blankets and household effects to add to his comfort in the spirit-land, are entombed with the remains.

Another occasion for indulging in the potlatch is when some one is desirous of securing extraordinary influence in his tribe, generally a chief seeking to establish superior position or popularity over some rival. Natives have been known to save their means for years, augmenting them by industry and self-denial, in order finally to give a grand and unequaled feast of this character. When the time arrives not only are all the host's own tribe invited, but those of the next nearest tribes not akin to his own. Such a festival often lasts for a whole week, until the last blanket of the giver is sacrificed. These strange festivals, we were told, are fast passing into disuse, at least among those tribes brought most in contact with the whites, though on a smaller scale they do still exist all over the southern region of Alaska.

There is, perhaps, no positive evidence that cannibalism ever prevailed among the Indians of this region, yet it is gravely hinted that it did on the occasion of these funeral potlatches years ago. To sacrifice the life of one or more of the slaves of the deceased we know was common, and if their bodies were not barbecued and eaten, then these natives of the North Pacific were entirely different in this respect from those who lived in the

South Pacific. The medicine-men, even to-day, devour portions of corpses, believing that they acquire control of the spirit of the deceased thereby, and gain influence over demon spirits in the other sphere. Such practices are, however, rare, though Mr. Duncan of Metla-katla tells us he has witnessed the repulsive performance. The places near each hamlet where the dead are finally placed often number many more graves, or square boxes containing the bodies, than there are present inhabitants in the settlement. All this region was formerly many times more populous than it is to-day. Here, as in Africa, New Zealand, California, and Australia, where the white man appears permanently, the black man slowly but surely vanishes. The progress of civilization, as we call it, is fatal to native, savage races all over the world. Catlin, who lived among and wrote so well about our Western Indians, summed up the matter thus: " White man — whiskey — tomahawks — scalping-knives — guns, powder and ball — smallpox, debauchery — extermination." But it is not alone gunpowder, rum, and lasciviousness which are the active agents to this end; there is also a subtle influence which is not clearly understood, and which it is difficult to define, but which is as potent, if not more so, than the agencies above suggested. The destiny which heaven decrees for a people will surely come to them. This has been clearly exemplified in the instance of the North American Indians, as well as among the South Sea Islanders in Australia and the Hawaiian Islands. Of an entire and intelligent people, the aborigines who once occupied Tasmania, there is not to-day a living representative! The land is solely possessed and occupied by white Europeans, before whom the natives have steadily vanished like dew before the sun.

Mr. Frederick Whymper, who wrote about the Northwest some twenty years ago, speaking upon this subject, refers to the experience of a Mr. Sproat, a resident of the region near Puget Sound, who employed large numbers of natives as well as whites in manufacturing lumber. Mr. Sproat conducted his large business and the place where it was established on temperance principles; no violence or oppression of any sort was permitted towards the natives. They were in fact better fed, better clothed, and better taught than they had ever been before. It was only after a considerable time that any symptom of a change was observed among the Indians. By and by a listlessness seemed to creep over them, and they "brooded over silent thoughts." At first they were surprised and bewildered by the presence of the white men, and the machinery and steam vessels which they brought with them. They seemed slowly to acquire a distrust of themselves, and abandoned their old practices and tribal habits, until at last it was discovered that a higher death-rate was prevailing among them. "No one molested them," says Mr. Sproat; "they had ample sustenance and shelter for the support of life, yet the people decayed. The steady brightness of civilized life seemed to dim and extinguish the flickering light of savageism, as the rays of the sun put out a common fire."

Upon the same subject and people, H. W. Elliott says: "These savages were created for the wild surroundings of their existence; expressly fitted for it, and they live happily in it; change the order of their life, and at once they disappear, as do the indigenous herbs and game before the cultivation of the soil and the domestication of animals." We shall not comment upon these remarks, though to us it is an extremely interesting subject; the reader must draw his own inference.

The men of these native tribes are strong and vigorous; the women are, however, forced to perform most of the domestic labor, and all of the drudgery, yet it was observed that they held the purse strings. That is to say, a native buck always defers to his wife in any matter of trade as to the price either to ask or to pay. The women of Alaska are certainly in a better condition and are better treated than those belonging to any of our Western Indian tribes, with whom we are acquainted. Though they are called upon to do much menial work, they do not seem to be actually abused. The male Alaskan performs a certain liberal share of domestic duties, but not so with the Indian of our Western reservations. The latter makes his wife a beast of burden. They are generally clothed in the garments of civilization, though of coarse material and of the cheapest manufacture. The ready-made clothing store has reached even the islands of the North Pacific. Polygamy is common among the aborigines, chastity is little heeded, and young girls are sold by their mothers for a few blankets, she and not the father having the acknowledged right of disposing of them. Dr. Sheldon Jackson writes most feelingly as follows: "Despised by their fathers, sold by their mothers, imposed upon by their brothers, and ill-treated by their husbands, cast out in their widowhood, living lives of toil and low sensual pleasure, untaught and uncared for, with no true enjoyment in this world and no hope for the world to come, crushed by a cruel heathenism, it is no wonder that many of them end their misery and wretchedness by suicide."

It was found on inquiry that the ratio of births among the Alaskan shore tribes was considerably greater than among civilized communities, but the death-rate is, on the other hand, excessive. The wretched ignorance of the mothers as to the observance of the simplest sanitary laws, as well as the gross exposure of their infants, is the principal cause of this needless mortality.

The aborigines, where not brought in contact with the government schools and missionaries, still retain their system of fetich worship, being very much under control of their medicinemen, who pretend to influence the demons of the spirit world, so feared by the average savage. Their moral degradation is extreme, and their practices in too many instances are terrible to relate. Slaves are sacrificed, as already stated, at the owner's death, that they may go before and prepare for his arrival in the future state. Vile witchcraft is still believed in among most of the tribes, and murderous consequences follow in many cases. All kinds of barbarity are inflicted upon women, children, and slaves. We are told by Dr. Sheldon Jackson that it was surprising to see how quickly these savage practices yielded to the power of Christian teachings, and how rapidly they faded away before the influence of association with a few intelligent, conscientious white teachers. What these people need is education and Christian influence, which will work a great and rapid reform among them in a single generation.

The canoes of the tribes about the Alexander Archipelago are dug out of well-chosen cedar logs, and are given the really fine lines for which they are remarkable by means of hot water and steam, together with the use of cunningly devised braces and clamps. The wood being once thoroughly dried in the desired shape, will retain it. Wondering how the exquisite smoothness was produced in forming their boats without a carpenter's plane, it was found by inquiry that the natives dry the coarse skin of the dogfish and use it as we do sandpaper. The time spent upon the construction and ornamentation of these canoes is apparently of no consideration to the native, and the market value of the best will average one hundred dollars. It is the Alaskan's most necessary and most prized piece of property. Some which we saw were eighty feet in length, and capable of holding one hundred men. It must be remembered that almost the entire population live on the coast or river banks in a country where there are no roads. These canoes have no seats in them; the rower places himself on the bottom, and thus situated uses his paddles with great dexterity. They are quite unmanageable by a white man who is not accustomed to them, as much so at least as a birch canoe, such as the Eastern Indians build on the coast of Maine. But the Alaskan boat is far superior to the birch-bark canoe in every respect. We saw one paddled by a boy at Pyramid Harbor, neat and new, which the lad, say twelve years of age, had dug out of a spruce log with his own hands, quite unaided. Its lines were admirable, and the finish was excellent. When the sun beats down upon these boats, the owner splashes water upon the sides about him to prevent their warping, and for this purpose carries a thin wooden scoop. When not in use they are carefully covered up to shelter them from the sun's rays. Some tribes use a double paddle, that is, an oar with a blade at each end, which they dip on one side and the other alternately; other tribes use the single-bladed paddle. Each one of the males among the natives has his canoe, for the water is his only highway, and without his boat he would be as helpless as one of our Western Indians on the plains without his pony. When the "dug-outs" are drawn up upon the shore in scores, they present a curious appearance, packed with grass and covered with matting to keep them from being cracked and warped by the sun. The bows and stern of many of them are elaborately carved totem-fashion, and also painted in strange designs with a black pigment. The fore part of the boat rises with an upward sheer, and is higher at the prow than at the stern. There is another form of boat used by the Eskimos and natives of the outlying islands, being a simple frame of wood, covered with sea-lion skin from which the hair has been removed. These boats are covered over the tops as well as the bottoms, being almost level with the sea, leaving only a hole for the occupant to sit in, thus making them absolutely watertight, a life-boat, in fact, which will float in any water so long as they will hold together. The waves may dash over them but cannot enter them. These skin-covered boats, admirably adapted to their legitimate purpose, are known on the coast as "bidarkas," in the management of which the natives evince great skill, making long journeys in them, and braving all sorts of weather. Like the Madras surf-boats, no nails are used in their construction, either in the skeleton frame or in putting on the covering, the several parts being lashed and sewed together in the most artistic fashion with sinews and leather thongs, which enables them to bear a greater strain than if they were held together by any other means. The thongs admit of a certain degree of flexibility when it is required, an effect which cannot be got with nail fastenings.

CHAPTER XV.

Still sailing Northward. — Multitudes of Water-Fowls. — Native Graveyards. — Curious Totem-Poles. — Tribal and Family Emblems. — Division of the Tribes. — Whence the Race came. — A Clew to their Origin. — The Northern Eskimos. — A Remarkable Museum of Aleutian Antiquities. — Jade Mountain. — The Art of Carving. — Long Days. — Aborigines of the Yukon Valley. — Their Customs.

STILL sailing northward, large numbers of ebon-hued cormorants are seen feeding on the low, kelp-covered rocks, contrasting with the snowy whiteness of the gulls. Big flocks of snipe, ducks, and other aquatic birds line the water's edge, or rise in clouds from some sheltered nook to settle again in our wake. Higher up in air a huge bald-headed eagle is in sight nearly all the while, as we sail along the winding watercourse. The eagles of Alaska, unlike those of other sections of the globe, are not a solitary bird, but congregate in considerable numbers, and residents told us they had seen a score of them roosting together on the branches of the same tree, but we must confess to never having seen even two together. Elsewhere the eagle is certainly a bird whose solitary habits are one of its marked characteristics. We observe here and there near native villages, more square boxes and totem-poles indicating the resting-places of the dead. Some tribes continue to burn their dead, and these boxes contain only the ashes, but the missionaries and the whites generally have so opposed the idea of cremation that many of the natives have abandoned it. The burial above-ground in the square boxes referred to is a peculiar idea. These coffins, if they may be so called, are about three feet and a half long by two and a half wide, and are often elaborately carved and painted with grotesque figures. The corpse is disjointed and doubled up in order to get it into this compass, though why this is done when a longer box would so much simplify matters, no one seems to know. We were told that some of the Alaskan tribes used to place their dead in trees, or on the top of four raised poles, a similar practice to that which once prevailed among certain tribes of our Western Indians, but the mode just described is that which most generally prevails. There seems to be some difference of opinion as regards the real significance of the totem-poles. They appear to be designed in part to commemorate certain deeds in the lives of the departed, near whose grave they are reared, as well as to indicate the family arms of those for whom they are erected. Thus, on seeing one special totem-post surmounted by a wolf carved in wood, beneath which a useless gun was lashed, inquiry was made as to its significance, whereupon we were told that the deceased by whose grave it stood had been killed while hunting wolves in the forest. This was certainly a very literal way of recording the fate of the hunter.

Some tribes adopt the crow, some the hawk, and some the bear or the whale, as their distinctive tribal emblem. The poles are carved from bottom to top, averaging thirty or forty feet in height, — though some are nearly a hundred feet high, — and from three to four feet in diameter, the height also signifying the importance of the individual, that is, his social grade or standing in the tribe. Some of the carvings are mythological, for these people have an oral mythology of the most fabulous character, which has been handed down from father to son for many centuries. The carvings on the coffin-boxes, though often elaborate, to a white man's eye are meaningless. As we have said, when a chief dies, some valuable personal effects are always deposited with his body in the coffin, and one would suppose that such objects were safe from pilfering fingers of even strangers; yet these articles are constantly offered for sale, and are eagerly purchased by curio-hunters who come hither from various parts of this country.

The aborigines of Alaska are divided into various sub-tribes, such as Hooniahs, Tongas, Auks, Kasa-ans, Haidas, Sitkas, Chinooks, Chilcats, and so on.

Ivan Petroff, who was sent by the United States Government to Alaska in 1880, as special agent of the census, divides the native population of the Territory as follows: —

FIRST. — The Innuit or Eskimo race, which predominates in numbers and covers the littoral margin of all Alaska from the British boundary on the Arctic to Norton Sound, the Lower Yukon, and Kuskoquin, Bristol Bay, the Alaska Peninsula, Kodiak Island, mixing in, also, at Prince William Sound.

SECOND. — The Indians proper spread over the vast interior in the north, reaching down to the seaboard at Cook's Inlet and the mouth of the Copper River, and lining the coast from Mount St. Elias southward to the boundary and peopling the Alexander Archipelago.

THIRD. — The Aleutian race, extending from the Shumagin Islands westward to Attoo, — the Ultima Thule of this country, — whom Petroff terms the Christian inhabitants. These last certainly conform most fully to all the outward practices of civilization and universally recognize the Greek Church.

Whence these people originally came is a question which is constantly discussed, but which is still an unsolved problem. Some words in their language seem to indicate a Japanese origin, and some seem clearly to be derived from the Aztec tongue belonging to that peculiar people of the south. Hon. James G. Swain of Port Townsend, who has given years of study to the subject of ethnology as connected with the tribes of the Northwest, states that he found among them a tradition of the Great Spirit similar to that of the Aztecs, and that when he exhibited to members of the Haida tribe sketches of Aztec carvings, they at once recognized and understood them.

Copper images and relics found in their possession were identical with exhumed relics brought from Guatemala. These are certainly very significant facts, if not convincing ones. The Alaska natives have some Apache words in their language, which points to a common origin with our North American Indian tribes, but these suggestions are purely speculative. There are able students of ethnology who insist upon the origin of these Alaskans being Asiatic for various good and sufficient reasons, instancing not only their personal appearance, but the similarity of their traditions and customs to those of the people of Asia. To have come thence it is remembered that they had only to cross a narrow piece of water forty miles wide. This passage is frequently made in our times by open boats. At certain seasons of the year, though in so northern a latitude, the strait is by no means rough. Mr. Seward says: "I have mingled freely with the multifarious population, the Tongas, the Stickeens, the Kakes, the Haidas, the Sitkas, the Kootnoos, and the Chilcats. Climate and other circumstances have indeed produced some differences of manners and customs between the Aleuts, the Koloschians, and the interior continental tribes, but all of them are manifestly of Mongol origin. Although they have preserved no common traditions, all alike indulge in tastes, wear a physiognomy, and are imbued with sentiments peculiarly noticed in China and Japan."

The Eskimos proper differ but little from the southern and inland tribes of Alaska generally;

few of them are ever seen south of Norton Sound or the mouths of the Yukon. Their home is in the Arctic portion of the Territory, bordering the Frozen Ocean and Behring Strait. It is obvious that climatic influences create among them different manners and customs, causing also a slightly different physical formation, but otherwise they seem to be of the same race as the people of the Alaska Peninsula, the Aleutian Islands, or indeed of any of the several groups and of the mainland lying to the south. That these Eskimos resemble physically the Norwegian Lapps, to be met with at about the same latitude in the eastern hemisphere, is very obvious to one who has carefully observed both races in their homes. This similarity extends in rather a remarkable degree also to their dress as well as domestic habits.

In the region they occupy, near the source of the Kowak River, which empties into Kotzebue Sound by several mouths after a course of two or three hundred miles, is Jade Mountain, composed, as far as is known, of a light green stone which gives it the name it bears. An exploring party from the United States steamer Corwin brought away one or two hundred pounds of the mineral in the summer of 1884. The hardness and tenacity of these specimens are said to have been remarkable, as well as the exquisite polish which they exhibited when treated by the lapidist. Jade Mountain must be in latitude 68° north, between two and three hundred miles south of the Yukon above the line of Behring Strait.

Yet the exploring party found the thermometer to register 90° Fah. in the shade, while their greatest annoyance was caused by the mosquitoes. The Kowak abounds in salmon, pike, and white-fish. "The 'color' of gold," says the printed report of the expedition, " was obtained almost everywhere." Nearly eighty species of birds were collected, though the party were absent from the Corwin but about seven weeks. The white spruce was found to be the largest and most abundant tree, and the inhabitants all Eskimos.

The remarkable museum of ancient arms, dresses, wooden and skin armor, and domestic utensils exhibited in New York city in 1868 by Mr. Edward G. Fast, and which was collected by him while in the employment of our government among the people of the Northwest, revealed some very important facts as to their history. The collection proved clearly that two or three hundred years ago these natives of Alaska enjoyed a much higher degree of civilization than is exhibited by their descendants to-day. That they have deteriorated in industry, steadiness, and ability generally is obvious. The art of forging must have been known to them in the earlier times, as shown in this collection of admirable weapons, clearly of native manufacture and of most excellent finish. The art of carving was possessed by them in far greater perfection than they exhibit in our day, while the skillfully made dresses of tanned leather worn by the ancient Aleuts nearly equal those in which the warriors were clad who accompanied

Cortez and Pizarro when they landed on this continent. Mr. Fast was singularly fortunate in securing whole suits of armor, masks, and war implements for his unique museum of Alaskan antiquities. In association with Russians and Americans for a century, more or less, these aborigines have readily adopted the vices of civilization, so to speak, and have sacrificed most of their own better qualities. Indolence generally has taken the place of the warlike habits and steadiness of purpose which must have characterized them as a people to a large degree before the whites came with firearms and fire-water. How forcibly is the law of mutability impressed upon us! From a state of comparative power and importance, this people has dwindled to a condition simply foreshadowing oblivion.

Rev. W. W. Kirby, a missionary who reached the valley of the Yukon by way of British Columbia, fully describes the Eskimos whom he mingled with in the northwestern part of the Territory. He considers them to be more intelligent than the average Alaska Indians, and far superior to them in physical appearance, the women especially being much fairer and more pleasing to look upon. They are more addicted to the use of tobacco than are these southern tribes, often smoking to great excess, and in the most peculiar manner, swallowing every swiff from their pipes, until they become so poisoned as to fall senseless upon the ground, where they remain in this condition for ten or fifteen minutes. They dress very neatly with deerskins, wearing the hair on the outside. The men have heavy beards, shave the crown of their heads, leaving the sides and back growth to fall freely about the face and neck. Mr. Kirby is obliged to censure the thievish propensities of this people, which was a source of great trouble and considerable loss to him. Speaking of his high northern latitude when among the Eskimos, he says: "As we advanced farther northward, the sun did not leave us at all. Frequently did I see him describe a complete circle in the heavens."

As far south as Pyramid Harbor, latitude 59° 11' north, the sun does not set in midsummer until about two o'clock in the morning, rising again four hours later. Even during these four sunless hours fine print can be read on the ship's deck without the aid of any other than the natural light.

Mr Kirby found the Indians of the Yukon valley to be rather a fierce and turbulent people, more like our Western Indians than any other tribes whom he met. Their country is in and about latitude 65° north, and beginning at the Mackenzie River, in British Columbia, runs through Alaska to Behring Strait. They were formerly very numerous, but have frequently been at war with the Eskimos north of them, and have thus been sadly reduced in numbers, though they are still a strong and powerful people.

There is a singular system of social division recognized among them, termed respectively Chit-sa, Nate-sa, and Tanges-at-sa, faintly representing the idea of aristocracy, the middle class, and the poorer order of our civilization. There is another peculiarity in this connection, it being the rule for a man not to marry in his own, but to take a wife from either of the other classes. Thus a Chit-sa gentleman will marry a Tanges-at-sa peasant without hesitation; the offspring in every case belonging to the class to which the mother is related. This arrangement has had a most beneficial effect in allaying the deadly feuds formerly so frequent among neighboring tribes, and which have been the cause of so reducing their memorial strength by sanguinary conflicts.

CHAPTER XVI.

Fort Wrangel. — Plenty of Wild Game. — Natives do not care for Soldiers, but have a Wholesome Fear of Gunboats. — Mode of Trading. — Girls' School and Home. — A Deadly Tragedy. — Native Jewelry and Carving. — No Totem-Poles for Sale. — Missionary Enterprises. — Progress in Educating Natives. — Various Denominations Engaged in the Missionary Work.

WE prefer to think it was to see the sun rise that we got up so early on arriving at Fort Wrangel, and not because of the torturing fact that our berth was too short at both ends, and kept us in a chronic state of wakefulness and cramp. The distance passed over in coming hither from Victoria was about eight hundred miles. The place, having about five hundred inhabitants, is advantageously situated on an island at the mouth of the Stickeen River, which rises in British Columbia and has a length of nearly two hundred and fifty miles. There is here an excellent and capacious harbor, surrounded by grand mountains, while lofty snow-crowned summits more inland break the sky-line in nearly all directions, — mountain towering above mountain, until the view is lost among far-away peaks, blue and indistinct. This elevated district contains wild goats, with now and then a grizzly bear, fiercest of his tribe, while in its ravines and valleys the little mule-deer, the brown bear, the fox, the land-otter, the mink, and various other animals abound. As to the small streams and river courses which thread the territory, they are, as all over this country, crowded with fish, the salmon prevailing. The inland haunts within twenty leagues of the coast are little disturbed by the natives. The abundance of halibut, cod, and salmon at their very doors, as it were, is quite sufficient to satisfy the demands of nature, and it is only when tempted by the white man's gold that the aborigines will leave the coast to go inland in search of pelts and meat, in the form of venison, goat, or bear flesh.

The town, consisting of a hundred houses and more, is spread along the shore at the base of a thickly wooded hill, flanked on either side by a long line of low, square, rough-hewn native cabins. A peep into the interior of these was by no means reassuring. Dirt, degradation, and abundance were combined. The few domestic utensils seen appeared never to have been washed, being thick with grease, while the stench that saluted the olfactories was sickening. There were no chairs, stools, or benches, the men and women sitting upon their haunches, a position which would be a severe trial to a white and afford no rest whatever, but which is the universal mode of sitting adopted by savage races in all parts of the world. The place was named after Baron Wrangel, governor of Russian America at the time when it was first settled, in 1834, being then merely a stockade post. After the United States came into possession of the country it was for a short time occupied by our soldiers, but ere long ceased to be held as a military post, the soldiers being withdrawn altogether from the Territory. It was soon discovered that the natives cared nothing for the soldiers; they could always get away from them in any exigency by means of their canoes; but they had, and still have, a wholesome fear of a revenue cutter or a gunboat, which can destroy one of their villages, if necessary, in a few minutes.

A steamer can always move very rapidly from place to place among the islands, making her presence felt without delay, when and where it is most needed. At the outset of our taking possession of Alaska, an example of decision and power was necessary to put the natives in proper awe of the government, and it followed quickly upon an unprovoked outrage committed by the aborigines. One of their villages, not far from Sitka, was promptly shelled and destroyed in half an hour. Since then there has been no trouble of consequence with any of the tribes, who have profound respect for the strong arm, and to speak plainly, like most savage races, for nothing else.

Fort Wrangel has two or three large stores for the sale of goods to the natives, and for the purchase of furs, Indian curiosities, and the like. It is also the headquarters of the gold miners, who gather here when the season is no longer fit for out-of-door work at the placers.

Seeing the natives crowding the stores, it was natural to suppose the traders were driving a good business, but a proprietor explained that these people were slow buyers, making him many calls before purchasing. They look an article over three or four different times before concluding they want it; then its cost is to be considered. The native's squaw comes and approves or disapproves; the article is discussed with the men's neighbors, and, finally, his resolution having culminated, he goes away to earn the money with which to make the purchase! "Such customers are very trying to our patience," remarked the trader, "but after you once understand their peculiarities it is easy enough to get along with them."

A truly charitable enterprise has been established here; we refer to the Indian Girls' School and Home, supported by the American Board of Missions, where the pupils are taught industrial duties appertaining to the domestic associations of their sex, as well as the ordinary branches of a common school education. No effort, we were told, is made to enforce any special tenets of faith, but these girls are taught morality, which is practical religion. The example is much needed here, both among these native people and the whites.

To show what strict adherents these Alaskans are to tribal conventionalities, we can do no better than relate a singular occurrence, for the truth of which Dr. Jackson is our authority.

"Near the Hoonah Mission, a short time ago, a deadly tragedy took place. A stalwart native came into the village and imbibed too freely of hoochinoo. Walking along the street he saw a young married girl with whom he was greatly infatuated. The girl was afraid to meet him and turning ran to her house. The man gave pursuit and gained entrance to the house. All the inmates escaped in terror. The desperado boldly continued his hunt for the woman, and the husband of the woman with a few friends took refuge in his own house again. The ravishing fiend returned, and demanding admittance battered in the door with an axe, and as he entered was shot and instantly killed. The friends of the dead man met in council, and according to their custom demanded a life for his life. The husband and protector of his wife's virtue gave himself into the custody of his enemies and was unceremoniously killed!"

The production of native jewelry is a specialty here, and some of the silver ornaments of Indian manufacture are really very fine, exhibiting great skill and originality, if not refined taste. Their carvings in ivory are exceedingly curious, skillful, and attractive, especially upon walrus teeth, whereon they will imitate precisely any pattern that is given to them, with a patient fidelity equaling the Chinese. The native designs are far the most desirable, however, being not only typical of the people and locality, but original and fitting. The time devoted to a piece of work seems to be of no consideration to a native, and forms no criterion as regards the price demanded for it. From the sale of these fancy articles the aborigines receive annually a considerable sum of money. It is indeed surprising how they can get such results without better tools. With some artistic instruction they would be capable of producing designs and combinations of a choice character, and which would command a market among the most fastidious purchasers. Their present somewhat rude ornaments have attracted so much attention that two or three stores in San Francisco keep a variety of them for sale. But it is the charm of having purchased such souvenirs on the spot which forms half their value.

Speaking of these souvenirs, the author was shown some stone carvings at Victoria, on the passage from Puget Sound northward, which were of native manufacture, and thought to be idols. It was afterwards learned that these were the works of the Haidas of Queen Charlotte Island, about seventy or eighty miles north of Vancouver Island. There is here a slate-stone, quite soft when first quarried, which is easily carved into any design or fanciful figure, but which rapidly hardens on exposure to the air. The stone is oiled when the carving is completed, and this gives it the appearance of age, as well as makes it dark and smooth. The natives of this northwest coast do not worship idols, therefore these are not objects of that character, though they are curious and interesting. It is among these Haidas that the practice of tattooing most prevails, and they still cover their bodies with designs of birds, fishes, and animals, some of which are most hideous caricatures. This tribe is said to be the most addicted to gambling of any on the coast, the demoralizing effect of which is to be seen in various forms among them.

Fort Wrangel has several demon-like totem-poles. There is a sort of fascination attached to these awkward objects which leads one carefully to examine and constantly to talk about them. Before some cabins there are two of the weird things, covered with devices representing both the male and female branches of the family which occupies the cabin. It was found that much more importance was attached to these emblems here than had been manifested farther south. An interested excursionist who came up on our steamer, wishing to possess himself of a totem-pole, found one at last of suitable size for transportation, and tried to purchase it, but discovered that no possible sum which he could offer would be considered as an equivalent for it. All of his subsequent efforts in this line proved equally unsuccessful so far as totem-poles were concerned, and yet we remember that they are to be found in many of our public museums throughout the States, and we have seen large ones lying upon the ground moss covered and neglected. It appeared to be only the rich native who indulged in an individual totem-pole. The cost of one, say forty or fifty feet long, carved after the orthodox fashion, with the free feast given at all such raisings, is said to be over a thousand dollars. The more lavish the expenditure on these occasions, the greater the honor achieved by the host.

There is a successful day-school established here

besides the Indian Girls' Home, which is accomplishing much good in educating the rising generation, and in introducing civilized manners and customs. The children evince a fair degree of natural aptitude, learning easily to read and write, but are a little dull, we were told, in arithmetic. Adult, uneducated natives, however, are quick enough at making all necessary calculations in their trades with the whites, either as purchasers of domestic goods, or in selling their peltries. The Presbyterians, Methodists, Episcopalians, Moravians, Quakers, Baptists, and Roman Catholics all have missionary stations in different parts of the country. Schools have also been established for the general instruction of whites and natives at Juneau, Sitka, Wrangel, Jackson, and other localities under direction of our government officials, and proper teachers have been supplied, the whole system being under the supervision of a competent head. Mrs. J. G. Hyde, who teaches school at Juneau, in her last year's report, says: "Many of the scholars, who, when the term began last September, could not speak a word of English, can now not only speak, but read and write it. They can also spell correctly and are beginning in the first principles of arithmetic. To the casual observer perhaps nothing seems more absurd than the attempt by any process to enlighten the clouded intellect of this benighted people. Indeed, the most squalid street Arabs might be considered a thousand times more desirable as pupils. But a few days' work among and for them convinces the teacher that she has not a boisterous, uncontrollable lot of children, but as much the opposite as it is possible to imagine. Children who habitually refrain from playing during intermission that they may learn some lesson or how to do some fancy work are not to be classed with the wild, wayward, or vicious. Boys who, when their regular lessons are done, are continually designing and drawing cannot be said to be entirely devoid of talent worthy of cultivation. While the development must be slow in most cases, there are a few who would compare favorably with white children. Their abnormal development of the faculty of form gives them an inestimable advantage over their more favored pale-face brothers in acquiring the art of writing and drawing. Their mind acts very slowly, but they make up in tenacity of purpose what they lack in aptness."

At Sitka there is an industrial school which is very successful training native boys and girls in mechanical and domestic occupations, and of which we will speak in detail in a further chapter.

CHAPTER XVII.

Schools in Alaska. — Natives Ambitious to learn. — Wild Flowers. — Native Grasses. — Boat Racing. — Avaricious Natives. — The Candle Fish. — Gold Mines Inland. — Chinese Gold-Diggers. — A Ledge of Garnets. — Belief in Omens. — More Schools required. — The Pestiferous Mosquito. — Mosquitoes and Bears. — Alaskan Fjords. — The Patterson Glacier.

THE general plan of this school at Wrangel struck us as being the most promising means of improvement that could possibly be devised and carried forward among the aborigines of Alaska. We were informed that fourteen government day schools were in operation in the Territory, under the able supervision of that true philanthropist, Dr. Sheldon Jackson, United States General Agent for Education in the Territory. The natives almost universally welcome and gladly improve the advantages afforded them for instruction, especially as regards their children. Many individual cases with which the author became acquainted were of much more than ordinary interest; indeed, it was quite touching to observe the eagerness of young natives to gain intellectual culture. Surely such incentive is worthy of all encouragement. One could not but contrast the earnestness of these untutored aborigines to make the most of every opportunity for learning with the neglected opportunities of eight tenths of our pampered children of civilization. Here is the true field of missionary work, the work of education.

In the neighborhood of Fort Wrangel plenty of sweet wild flowers were observed in bloom, some especially of Alpine character were very interesting, — "wee, modest, crimson-tipped flowers," — while the tall blueberry bushes were crowded with wholesome and appetizing fruit, with here and there clusters of the luscious salmon-berry, yellow as gold, and so ripe as to melt in the mouth. At the earliest advent of spring the flowers burst forth in this latitude with surprising forwardness, a phenomenon also observable in northern Sweden and Norway. Such white clover heads are rarely seen anywhere else, large, well spread, and fragrant as pinks. Among the ferns was an abundance of the tiny-leaved maiden's hair species, with delicate, chocolate stems. The soil also abounds in well-developed grasses, timothy growing here to four feet and over in height, and the nutritious, stocky blue grass even higher. Vegetation during the brief summer season runs riot, and makes the most of its opportunity. Although south of Sitka, Fort Wrangel is colder in winter and warmer in summer, on account of its distance from the influence of the thermal ocean current already described.

Sometimes a purse is made up among the visitors here and offered as a prize to the natives in boat-racing. A number of long canoes, each with an Indian crew of from ten to sixteen, take part in the aquatic struggle, which proves very amusing, not to say exciting. The native boats are flat-bottomed, and glide over the surface of the water with the least possible displacement. An Alaskan is seen at his best when acting as a boatman; he takes instinctively to the paddle from his earliest youth, and is never out of training for boat-service so long as he lives and is able to wield an oar. No university crew could successfully compete with these semi-civilized canoeists. Well-trained naval boat-crews have often been distanced by them.

The avariciousness of the natives is exhibited in their readiness to sell almost anything they possess for money, even to parting with their wives and daughters to the miners for base purposes; though, as we have seen, they do draw the line at totem-poles. It should be understood that these queerly carved posts are emblems mostly of the past; that is to say, although the natives carefully preserve those which now exist, few fresh ones are raised by them. Toy effigies representing these emblems are carved and offered for sale to curio-hunters at nearly all of the villages on the coast, and as a rule are readily disposed of.

There is very little if any use in Alaska for artificial light during the summer season, while nature's grand luminary is so sleepless; but when these aborigines do require a lamp for a special purpose, they have the most inexpensive and ingenious substitute ever ready at hand. The water supplies them with any quantity of the ulikon or candle-fish, about the size of our largest New England smelts, and which are full of oil. They are small in body, but over ten inches in length. They are prepared by a drying process and are stored away for use, serving both for food and for light. When a match is applied to one end of the dried ulikon, it will burn until the whole is quite consumed, clear and bright to the last, giving a light equal to three or four candles. So rich are these fishes in oil that alcohol will not preserve them, a discovery which was made in preparing specimens for the Smithsonian Institution. When the Indians of the interior visit the coast, as many of them do annually, they are sure to lay in a stock of candle-fish to take back with them for use in the long Arctic night. This fish runs at certain seasons of the year in great schools from the sea, invading the fresh-water rivers near their mouths, when the natives rake them on shore by the bushel and preserve them as described. When boiled they produce an oil which hardens like butter, and which the Alaskans eat as we do that article, with this important difference, that they prefer their oil-butter to be quite rancid before they consider it at its best, while civilized taste requires exactly the opposite condition, namely, perfect freshness. Putrid animal matter would certainly poison a white man, but the Alaskan Indians seem to thrive upon it.

Some inland districts, which are most easily reached from this point, are rich in gold-bearing quartz and placer mines, but especially in the latter. We were credibly informed that over three million dollars' worth of gold was shipped from here in a period of five years, though no really organized and persistent effort at mining had been made, or rather we should say no modern facilities had been employed in bringing about this result. The machinery for reducing gold-bearing quartz has not yet been carried far inland because of the great difficulty of transportation. Gold quartz ledges are numerous and quite undeveloped in the neighborhood of Wrangel. The well-known Cassiar mines are situated just over the Alaska boundary on the east side in British Columbia, but the gold discoveries in Alaska proper are proving so much more profitable that those of the Cassiar district have ceased to attract the miners. There is a curious fact connected with these deposits of the precious metal in the region approached by the way of Wrangel. In more than one instance, as reported by Captain White of the United States Revenue Service, placer gold, which is usually sought for in the dry beds of river courses and in low lands, is here found on the tops of mountains a thousand feet high, where the largest nuggets of the precious metal yet found in the Northwest have been obtained. Many of the lumps of pure gold picked up in this region have weighed thirty ounces and over. The idea of finding placer deposits on the tops of mountains is a novelty in gold prospecting.

The Stickeen River, which is the largest in the southern part of the Territory, has its mouth in the harbor of Fort Wrangel, discoloring the waters for a long distance with its chalk-like, frothy flow, a characteristic of all Alaska streams into which the waters of the snowy mountains and glaciers empty. The river is navigable for light-draft stern-wheel steamers to Glenora, a hundred and fifty miles from its mouth. After reaching this place, the way to the Cassiar mines is overland for an equal distance by a difficult mountain trail, it being necessary to transport all provisions and material on the backs of natives, who have learned to demand good pay for this laborious service. The interior upon this route is broken into a succession of sharply-defined mountains, separated by narrow and deep valleys, similar to the islands off the mainland. This is so decided a feature as to lead Mr. George Davidson of the United States Coast Survey to remark: "The topography of the Alexander Archipelago is a type of the interior. A submergence of the mountain region of the mainland would give a similar succession of islands, separated by deep narrow fjords." The sandy bed and banks of the Stickeen are heavily charged with particles of gold, ten dollars per day each being frequently realized by gangs of men who manipulate the same only in the most primitive fashion. Numbers of Chinamen availed themselves of this opportunity until they were expelled by both the whites and the natives. The poor "Heathen Chinee" is unwelcome everywhere outside of his own Celestial Empire, and yet close observation shows, as we have already said, that these Asiatics have more good qualities than the average foreigners who seek a home on our shores.

The scenery of the Stickeen River is pronounced by Professor Muir to be superb and grand beyond description. Three hundred glaciers are known to drain into its swift running waters, over one hundred of which are to be seen between Fort Wrangel and Glenora. Near the mouth of the river is the curious ledge of garnet crystals, which furnishes stones of considerable beauty and brilliancy, though not sufficiently clear to be used as gems. Choice pieces are secured by visitors as cabinet specimens, however, and can be had, if desired, by the bushel, at a trifling cost. They occur in a matrix of slate-like formation, some so large as to weigh two or three ounces, and diminishing from that size they are found as small as a pin-head. It requires three days of hard steaming against the current to ascend the river as far as Glenora from the mouth, whereas the same distance returning, down stream, has frequently been made in eight or ten hours. So necessarily rapid is the descent of the Stickeen as to make the downward trip quite hazardous, except in charge of a careful pilot. In the neighborhood of Fort Wrangel there are some very active boiling springs, which the natives utilize, as do the New Zealanders at Ohinemutu, by cooking their food in them.

In the crater of Goreloi, on Burned Island, is a vast boiling spring, or rather a boiling lake, which has never been intelligently described, and which is represented by those who have seen it to be unique. This strange body of water is eighteen miles in circumference. The natives are well supplied with legends relating to these remarkable natural phenomena, including the extinct and active volcanoes. Genii and dreaded spirits are supposed by them to dwell in the extinct volcanoes, and to make their homes in the mountain caves. They believe that good spirits will not harm them, and therefore do not address themselves to such, but the evil ones must by some active means be propitiated, and to them their sole attention is given, or, in other words, their religious ceremonies when analyzed are simply devil worship. All of the tribes, if we except the Aleuts, are held in abject fear by their conjurers or medicine-men, who seemed to us to be the most arrant knaves conceivable, not possessing one genuine quality to sustain their assumptions except that of bold effrontery. This seems particularly strange, as the aborigines of the Northwest are more than ordinarily intelligent, compared with other half-civilized races, both in this and other lands.

They are firm believers in signs and omens. When Rev. Mr. Willard and wife first came to the Chilcat country the winter was one of deep snows and stormy weather. The natives said that the weather-gods were angry at the new ways of the missionaries. A child had been buried instead of burned on the funeral pyre in accordance with their customs. The mother of the child became alarmed and felt that her life was in jeopardy for permitting her child to be buried, so she kindled a fire over the grave in order to appease the gods and bring fair weather. At school the children had played new games and mocked wild geese. So the girls of the Sitka Training School brought on a very cold spell of weather by playing a game called "cat's-back," and which caused a commotion at the native village. A white man out with some natives picked up some large clam-shells on the beach to bring home with him; the natives remonstrated with him, saying that "a big storm may overtake us, our canoe might capsize, and all be drowned the next time we go on the water."

In tempestuous weather the native propitiates the spirit of the storm by leaving a portion of tobacco in the rock-caves alongshore, but in calm weather he smokes the weed himself. It was noticed, however, that the aboriginal Alaskans were little given to the use of tobacco, less, indeed, than any semi-civilized race whom the writer has ever visited.

Governor Swineford, in his annual report to the department at Washington, dated 1886, says: "I have no reason to change or modify the estimate I had formed on very short acquaintance of the character of the native Alaskans. They are a very superior race intellectually as compared with the people generally known as North American Indians, and are as a rule industrious and provident, being wholly self-sustaining. They are shrewd and natural-born traders. Some are good carpenters, others are skillful workers in wood and metals. Not a few among them speak the English language, and some of the young men and women have learned to read and write, and nearly all are anxious for the education of their children."

Our government should act upon this hint and freely establish the means of education among the Alaskans. True, it is systematically engaged in promoting the cause in various ways, though not very energetically, Congress having voted forty-five thousand dollars to be expended for the purpose during the year 1889. "School-houses are the republican line of fortifications," said Horace Mann. "Among those best known," says Dr. Sheldon Jackson, speaking of the native tribes, "the highest ambition is to build American homes, possess American furniture, dress in American clothes, adopt the American style of living, and be American citizens. They ask no special favors from the American government, no annuities or help, but simply to be treated as other citizens, protected by the laws and courts, and in common with all others furnished with schools for their children." It was made the duty of the Secretary of the Interior, by the act providing a civil government for Alaska, to make needful and proper provision for the education of all children of school age without reference to race or color, and all true friends of progress and humanity will urge the matter until a common school is established in every native tribe and settlement having a sufficient number of children.

We were told that there is good hunting inland a short distance from Fort Wrangel; winter, however, is the only season when this can be successfully pursued near to the coast in the wild districts. The marshy " tundra " is then frozen and covered with snow, making it possible to cross. This is the period of the year also when the natives of the interior prosecute their most successful trapping and hunting, coming down to the coast by the river in the summer to sell their pelts and to purchase stores of the white traders. The Russians have long since taught the aborigines to depend much upon tea, but they care very little for coffee. Rifles are greatly prized by them, and though they are contraband nearly every Indian manages to possess one and knows how to use it most effectually. They are very economical of ammunition, and never throw away a shot by carelessness.

The pestiferous and ubiquitous mosquito is not absent from these high latitudes. They are very troublesome during the short summer season in northern Alaska as well as among the islands of the Alexander Archipelago. Strange that so frail an insect should have reached as far north as man has penetrated. Even while climbing the frosty glaciers the excursionist will find both hands required to prevent their biting his face from forehead to chin. If they are a persistent pest in equatorial latitudes, they are ten times more venomous and voracious in these regions during certain seasons. The author has experienced this fact also in Norway at even a much higher latitude than he visited in the western hemisphere. The bites of these mosquitoes fortunately, like all flesh wounds in this northern region, heal quickly, venomous as they are, owing to the liberally ozonized condition of the atmosphere as well as the absence of disease germs and organic dust.

It is said that when the otter hunters or others among the aborigines get wounded in any way, their treatment is simple and efficacious, and however severe the wound may be, it is nearly always quickly healed. The victim of the accident puts himself uncomplainingly on starvation diet, living upon an astonishingly small amount of food for a couple of weeks, and the cure follows rapidly.

Frederick Schwatka, in his excellent book entitled "Along Alaska's Great River," tells how the mosquitoes conquer and absolutely destroy the bears, and it seems that the native dogs are sometimes overcome by them in some exposed districts of the Yukon valley. The great brown bear, having exhausted the roots and berries on one mountain side, cross the valley to another range, or rather makes the attempt to do so, but is not always successful. Covered by a heavy coat of hair on his body, his eyes, nose, and ears are the only vulnerable points of attack for the mosquitoes, and hereon they congregate, surrounding the bear's head in clouds. As he reaches a swampy spot they increase in vigor and numbers, until the animal's forepaws become so occupied in striving to keep them off that he cannot walk. Then Bruin becomes enraged, and, bear-like, rises on his hind legs to fight. It is a mere question of time after this stage is reached until the bear's eyes become so swollen from the innumerable bites that he cannot see, and in a blind condition he wanders helplessly about until he gets mired and starves to death. The cinnamon and black bears are most common, the grizzly being less frequently met with. The great white polar bears are not found south of Behring Strait, though they are numerous on the borders of the Arctic Ocean.

At every landing made by the steamer on our meandering course among the islands Indians come to the wharves to offer their curios or home-made articles, only valuable as souvenirs of the visit. As they mass themselves here and there, either on the shore or the ship's deck, they form picturesque groups, made up of bucks, squaws, and papooses, presenting charming bits of color, while they amuse the stranger by their peculiar physiognomy and manners. During the excursion season they must reap quite a harvest by the sale of baskets and various domestic trinkets.

After leaving Fort Wrangel we are soon in the wild, picturesque, and sinuous narrows which bear the same name. The water is shallow; here and there are many dangerous rocks in the channels. Inlets or fjords are often passed, so quiet and inviting in their appearance as to tempt the traveler to diverge from the usual route. Some of these marine nooks are deep enough to float the largest ship, yet far down through the clear water one can see gardens of zoöphytes invaded by myriads of curiously shaped fish, large and small. The bottom of these waters, like the land and sea of Alaska, teems with animal life. A few hours' dredging would supply the most enthusiastic naturalist with ample material for a year's study. In the many stops of the steamer to take or deliver freight, brief boat excursions can be enjoyed. On one of these occasions we saw the first live octopus, or devil-fish, with two of its fatal arms encircling a small fish, which, after squeezing out its life, the octopus would devour. The one which was seen on this occasion was not very large, the rounded body being, perhaps, eighteen or twenty inches across, but its vicious looking tentacles, six in number, two of which securely clasped its victim, were each three times that length. The large eyes seemed out of proportion to the animal's size, and were placed on one side like those of the flounder.

The Patterson glacier is the first of the many which come into view on this part of the voyage, but they multiply rapidly as we steam northward. It is vast in proportions, though partly hidden behind the moraine which it has raised. Three or four miles back from its front rises a wall of solid ice nearly a thousand feet in height. The whole was rendered marvelously beautiful, lighted up as we saw it by bright noonday sunshine, which brought out its frosty and opaline colors of white, scarlet, and blue, in brilliant array. Little has been written about the Patterson glacier, but it is one of the most remarkable in size and other characteristics in all Alaska. Vessels from San Francisco have taken whole cargoes of ice from these Alaskan glaciers and transported the same for use in California. There seems to be no reason why the gathering of such a supply should not be both possible and profitable, though ice can now be so easily manufactured by artificial means.

The fact that these glaciers are slowly decreasing in size leads to the conclusion that the extreme Arctic temperature in the north is slowly growing to be less intense. Intelligent captains of whaleships have made careful observations to a like effect. It was once tropical in the Yukon valley, — of that there is evidence enough; who can say that it may not again be so a few thousand years hence?

CHAPTER XVIII.

Norwegian Scenery. — Lonely Navigation. — The Marvels of Takou Inlet. — Hundreds of Icebergs. — Home of the Frost King. — More Gold Deposits. — Snowstorm among the Peaks. — Juneau the Metropolis of Alaska. — Auk and Takou Indians. — Manners and Customs. — Spartan Habits. — Disposal of Widows. — Duels. — Sacrificing Slaves. — Hideous Customs still prevail.

BEFORE reaching Juneau we explored Takou Inlet, where there are two large glaciers, one with a moraine before its foot, the other reaching the deep water with its face, so as to discharge icebergs constantly. The bay was well filled with these, some of which were larger than our steamer (the Corona), and all were of such intense blue, mingled with dazzling white, as to recall the effect realized in the Blue Grotto of Capri. This berg-producing glacier was corrugated upon its surface in a remarkable manner, being utterly impassable to human feet. It was nearly a mile in width and its length indefinite; we doubt if it has ever been explored. A thousand ice and snow fed streams poured into the bay from the surrounding mountains, which completely walled in the broad sheet of water, so sprinkled with ice-sculpture in all manner of shapes. The ceaseless music of falling water was the only noise which broke the silence of the scene. A cavalcade of fleecy clouds, kindly forgetting to precipitate themselves in form of rain, floated over our heads, producing delicate lights and shades, with creeping shadows upon the surrounding mountains. The steamer's abrupt whistle was echoed with mocking hoarseness from the surrounding cliffs, causing the myriads of white-winged wild fowl to rise from the icebergs until the air was filled with them like snowflakes. How wonderful it was! A broad clear flood of sunshine enveloped the whole; everything seemed so serene, so grand, the sky so blue, and the angels so near. It was all as magnificent as a gorgeous dream, to the thoughtful observer a living poem. Close in to the precipitous cliffs of the myrtle-green hills were inky shadows, which formed the requisite contrast to the crystal clearness of the surroundings. For thousands of years this glacial action has been going on, the story of the earth is so old; but its beauty is ever young, its loveliness eternal.

On our way up Gastineau Channel — the tidewaters of which have a rise and fall of sixteen feet — we have presented to us veritable Norwegian scenery, under a pale amethyst sky fringed at the horizon with orange and crimson; now gliding close to precipitous cliffs enlivened by silvery streams leaping down their sides, and now passing the mouths of inlets winding among abrupt mountains leading no one knows whither, for there are no maps or charts of these lateral channels. The Indian canoes may have occasionally penetrated them, but never the keel of the white man. On the left stand the tall peaks of Douglas Island, and on the right the jagged Alps of the mainland, both rising to a height of a thousand feet or more, on the continent side backed by elevations still more lofty. The Takou River flows into the sea and gives its name to the neighborhood. Here the Hudson Bay Fur Company established and maintained a trading-post for several years. All this region is famous for its game, such as deer, bears, caribou, wolves, foxes, martens, and minks, together with the abounding big-horn sheep. In place of wool these latter have a coat somewhat like the red deer, and except in the size of their horns they resemble our domestic sheep. We are told that this district is also rich in gold placer mines, and according to Professor Muir it must eventually yield extremely profitable results to intelligent mining enterprise. In many localities the placers have paid for years, though worked by the most simple means. The experience of California will undoubtedly be repeated in Alaska; the great aggregate of gold which was realized there will be duplicated here. After due thought and personal observation relative to the subject, we are willing to stand or fall upon the correctness of this prediction. The result may not come in the next year, or that following, but it will come in the near future. Mining north of 54° 40′ is only in its infancy; its growth has been far more rapid, however, than it was at the south, both because of the richness of the mines, and because the business of mining is, and will continue to be, done more intelligently.

Just before reaching Juneau a singular phenomenon attracted our attention; it was a furious snowstorm among the mountain peaks, while all about us was quite calm and pleasant. The thick clouds of snow were driven hither and thither, from one pinnacle to another, writhing and twisting like a cyclone or water-spout at sea. It was a curious contrast, the storm raging in those far upper currents, while we enjoyed a gracious wealth of sunshine in a temperature of 65° Fah.

Juneau, located one hundred and fifty miles southeast of Sitka, and about three hundred north of Fort Wrangel, is already a considerable mining centre, with a population of about four thousand, situated not far from Takou district, and is the depot for the rich quartz and placer mines which are located in the region back of it. The site of the town is picturesque, being at the base of an abrupt mountain cliff which is decked with sparkling cascades. We were told that there is a rise and fall of twenty-four feet in the tide at the wharf of Juneau, but think perhaps eighteen feet would be nearer correct. The winter population is swelled by the influx of miners when the placers are not worked owing to snow and ice. Truth compels us to say that the residents here, of both sexes, are far from being of a desirable class. The Indians of this vicinity are of the Auk and Takou tribes; good traders and good hunters, but enemies of each other, though not given to open hostility. The native women, as if not content with the natural ugliness which has been liberally bestowed upon them by Providence, besmear their faces with a compound of seal-oil and lampblack, but for what possible reason, except that it is aboriginal Alaska fashion, one cannot divine. It is said that this is a sort of mourning for departed relations or friends; but the hilarity of those thus marked was anything but an indication of sorrow. We can well remember Yokohama wives, with blackened teeth and shaved eyebrows, who looked, if possible, a degree worse than these Alaskan women. In the latter case, however, the wives confessedly sought to make themselves hideous to prevent jealousy on the part of their husbands; but the native women here do not assign any plausible reason for smooching themselves in this offensive manner. When their faces are washed, a circumstance of rare occurrence, they are as white as the average of white people who are exposed to an out-of-door life. It is not the practice of the aborigines of either sex to wash themselves with water. They are sometimes seen to besmear their faces and hands with oil, which they carefully wipe off with a wisp of dry grass, or other substitute for the towel of civilization. The effect is to make the features shine like varnished mahogany; but as to cleanliness obtained by such a process, that does not follow.

If it were possible to discover a soap mine here there might be some hopes of introducing among the natives that condition which common acceptation places next to godliness. A traveling companion remarked that although milk and honey could not be said to flow in this neighborhood, oil does.

Many of the women, like those of the South Sea and the Malacca Straits, wear nose rings and glittering bracelets, while they go about with bare legs and feet. The author has seen all sorts of rude decorations employed by savage races, but never one which seemed quite so ridiculous or so deforming as the plug which many of these women of Alaska wear thrust through their under lips. The plug causes them to drool incessantly through the artificial aperture, though it is partially stopped by a piece of bone, ivory, or wood, formed like a large cuff-button, with a flat-spread portion inside to keep it in position. This practice is commenced in youth, the plug being increased in size as the wearer advances in age, so that when she becomes aged her lower lip is shockingly deformed. It is gratifying to be able to say that this custom is becoming less and less in use among the rising generation, and the same may be said as to tattooing the chin and cheeks. The hands and feet of the women are so small as to be noticeable in that respect.

The girls and boys endure great physical neglect in their youth, so that only the strongest are able to survive their childhood. It was surprising to see children of tender age of both sexes clothed only in a single cotton shirt, reaching to their knees, bare-legged, bare-footed, and bare-headed, yet apparently quite comfortable, while our woolen clothes and waterproofs were to us indispensable.

We were told that in infancy these children are dipped every morning into the sea, without regard to the temperature, or season of the year, commencing the operation when they are four weeks old. This heroic, Spartan treatment of the bath will probably harden, if it does not kill, but undoubtedly the latter result is the more likely of the two. The adults of some of the tribes break holes in the ice in midwinter, and bathe with marvelous fortitude, not for purposes of cleanliness, but declaring that it makes them "brave and strong, able to resist the cold, and to live long." The next hour, however, they may be found sitting on their hams as close to the fire in the middle of their unventilated cabins as they can get, closely wrapped in blankets, head and all. The prevalence among them of rheumatism and consumption shows that Nature cannot be outraged with impunity even by half-civilized Alaskans.

The natives do not seem to know anything about medicine, but when seriously ill they call in their shaman or medicine-man, and submit to his wild and senseless incantations, a process which would drive a civilized patient distracted. Fifty years ago an epidemic of small-pox swept away one third of the population of this part of the North Pacific coast, besides which, from various causes, the number in the several tribes is steadily decreasing. Vaccination having been introduced, a second visit of the dreaded disease just mentioned was accompanied with a very much smaller fatality. A scourge known as black measles is a frequent visitor among the youthful Alaskans, and is quite as fatal as small-pox.

Strong efforts are made by our government officials to keep intoxicating liquors out of the Territory, and the law makes them strictly contraband, but it is no more difficult or impossible to smuggle in Alaska than it is in New York or Boston. There are plenty of irresponsible whites ready to make money out of the aborigines. Rum is the native's bane, its effect upon him being singularly fatal; it maddens him, even slight intoxication means to him delirium and all its consequences, wild brutality and utter demoralization. Molasses is sold freely to them, and the Indians have learned how to distill rum from it, so that they secretly produce a vile and potent intoxicant, in spite of all prohibition.

When a native husband dies his brother's or sister's son, according to their custom, must marry the widow, but if there is no male relative of the husband's living, the widow may then choose for herself. If the individual who thus falls heir to a widow does not fancy the conditions, he must buy himself off, or fight the widow's nearest male relative. Oftentimes, if the new alliance is particularly disagreeable, the victim escapes by paying so much cash or so many blankets. There seems to be no hurt to a native's honor that pecuniary consideration will not promptly heal. Corporal punishment is considered by these aborigines to be a great disgrace, and is very seldom resorted to even with rebellious children. Theft is not looked upon as a crime; but if discovered, the thief must make ample restitution; and when his peculation is known he promptly does so without question or murmur. They have the duel as a decisive means of settling family feuds. When matters have come to the last resort, there is no secret about the matter. The two combatants fight publicly with knives, their friends looking on and singing songs while the combat lasts. But these duels, the same as with many other earlier savage practices, are now nearly obsolete. Like our Western Indians, their method of war was the ambush and surprise, and like them they scalped their prisoners and subjected them to savage cruelties. This also is more of the past than the present, as no open conflicts would now be permitted by the United States officials. The natives deck themselves with paint, — yellow ochre, — and look very much like the Sioux and Apache Indians in this respect. A century ago they were armed with flint-capped lances, bows, and arrows, but association with the whites has now supplied them with firearms. The old style of native weapons has consequently disappeared, except the lance with which they hunt the sea-otter. Firearms they do not use in this occupation, fearing to frighten away the valuable game altogether. They still manufacture bows and arrows for sale as curiosities to visiting strangers. They pride themselves upon their accomplishments in singing and dancing, but which to civilized ears and eyes are only the grossest caricatures. In these notes of the natives we refer to no one tribe, but to the aborigines of Alaska generally. The various tribes of course differ from each other. Those most in contact with the whites, having abolished many of their ancient habits, have adopted in a certain degree such customs as they see the white people follow. The holding of slaves is still practiced among them. Formerly, as we have said, one or two of these were sacrificed when their owner died, if he was a chief, in order that he might be well attended in the new sphere upon which he was entering; but this practice also has passed away. in most communities, with many other cruelties which were once common. These slaves are generally descendants of parents who were taken in battle during civil wars, though they are also bought and sold for so many otter-skins, or so many blankets. Such persons are always submissive, and accept the position in which they find themselves as a matter of course. This enforced servitude will soon be entirely abolished.

Female infanticide has not been uncommon with some tribes, but it does not prevail as has been represented by late writers. It is true that there have been cases where mothers, dreading to bring up their girls to such lives of hardship as they have themselves endured, have resorted to this desperate alternative, but careful inquiry did not satisfy us that such a practice now prevails if, indeed, it has not entirely ceased. In common with nearly all semi-civilized and savage races, **the native Alaskans regard their women more in** the light of slaves than as help-mates, and nearly all the hard work, except hunting and fishing, falls to their share. This is not a peculiarity of savage life, after all; horses and mules are not harder worked than are women in Germany and various parts of Europe. The writer has seen women carrying hods of bricks and mortar up long ladders in Munich, while their husbands drank huge "schooners" of beer and smoked tobacco in the nearest groggery.

Here and there among the several tribes, strange, unnatural, hideous customs are still extant, relative to wives about to become mothers, and as to young girls arriving at the age of puberty. We realize, however, that is not for us to look at this people through the lens of any small circumscribed moral code, but with kindly, hopeful views, guided by a due consideration of their normal condition. The conventionalities of civilization do not apply; latitude and longitude make broad differences as to what constitutes vice and virtue, reason or unreason. Modern instances are inadequate as a criterion of comparison. One who has traveled in many lands has learned to expand his horizon of judgment to accord with his geographical experience.

Notwithstanding the light in which the Alaskan regards his women, there seems to be a universal concession made to them in all matters of trade, wherein they undoubtedly hold the veto power, and in some other respects their domestic authority is promptly acknowledged. Just where the line is drawn does not seem to be clear to a stranger. After a native had sold us some trifle, his wife in more than one instance came and demanded it back again, carefully refunding the consideration which was given for the same. To this interference the husband seemed forced to submit in silence, — forced by the arbitrary custom of his tribe. We were told that even among themselves an agreement amounted to nothing at all, as they claim the right, and exercise it, of undoing any contract at will, provided the consideration which passed is promptly refunded. Even the white traders are obliged to yield to this singular idea to a certain extent, for the sake of peace.

The story so often told about polygamous wives, that is women with husbands in the plural, cannot be absolutely denied, but is an exaggeration of facts. Such relations we were told did exist, but to no great extent, among the tribes of Alaska.

CHAPTER XIX.

Aboriginal Dwellings. — Mastodons in Alaska. — Few Old People alive. — Abundance of Rain. — The Wonderful Treadwell Gold Mine. — Largest Quartz Crushing Mill in the World. — Inexhaustible Riches. — Other Gold Mines. — The Great Davidson Glacier. — Pyramid Harbor. — Native Frauds. — The Chilcats. — Mammoth Bear. — Salmon Canneries.

IN some portions of the country the aboriginal dwellings are constructed partly under ground; this is especially the case in the far north among the Eskimos proper, on the coast of the Polar Sea. Such cabins are entered by a tunnel ten feet long, so low and small as to compel the occupants to creep upon their hands and knees in passing through it. The tunnel-entrance, which always faces the most favorable point, is covered with a rude shed to protect it from the snow and the severity of the weather. The cabins are conical in form, covered with turf and mud, a hole being left at the top to permit the smoke to escape. The fire is built in the middle of the apartment on the ground. Around the space left for this purpose is a platform of a few inches in height arranged for living and sleeping upon. At night, in extreme cold weather, a flap of skins is so arranged that it can be drawn over the opening in the roof which serves as a chimney, and thus, the entrance being also closed, the occupants become hermetically sealed, as it were, thoroughly outraging all our modern ideas of ventilation. Twelve or fifteen persons are often found together in such a cabin with its one room, where the decencies of life are utterly ignored, and where the stench to civilized nostrils is really something dreadful to encounter.

This description refers to the winter homes of the people, where they hibernate like some species of wild animals, but for the milder portion of the year the Eskimos are nomadic, traveling hither and thither, seeking the most favorable locations for hunting and fishing, while living in rudely constructed camps. They use tents adapted for this itinerant life, made from prepared walrus hides supported by a light framework of wooden poles. The more thrifty supply themselves with canvas tents bought of the whites, as being handier for use and transportation.

Speaking of the interior of the country, we have the authority of Mr. C. F. Fowler, late agent of the Alaska Fur Company, and long resident in the country, and of Ex-Governor Swineford, both of whom have carefully investigated the subject, for stating that there exists a huge species of animals, believed to be representatives of the supposed extinct mammoth, found in herds not far from the headwaters of the Snake River, on the interior plateaus of Alaska. The natives call them "big-teeth" because of the size of their ivory tusks. Some of these, weighing over two hundred pounds each, were from animals so lately killed as to still have flesh upon them, and were purchased by Mr. Fowler, who brought them to the coast. These mammoths are represented to average twenty feet in height and over thirty feet in length, in many respects resembling elephants, the body being covered with long, coarse, reddish hairs. The eyes are larger, the ears smaller, and the trunk longer and more slender than those of the average elephant. The two tusks which Mr. Fowler brought away with him each measured fifteen feet in length.

The author has almost universally found among savage races at least a few very old people of both sexes, who were apparently revered and carefully provided for by their descendants and associates, but here among the aborigines aged persons are certainly not often to be seen. Whether it is that, hardy and robust as they generally appear to be, they do not, as a rule, live to advanced years, or that a summary method is adopted to get rid of them after they have outlived their usefulness, it is impossible to say. We were told that such is certainly the case with some of the tribes farthest from the influence and supervision of the whites, and that half a century ago the extremely old, being considered useless, were frequently " disposed " of. It is clear enough that there is nothing in the climate of this region in any way inimical to health and longevity.

The women of the Takou district are very expert and industrious. They occupy a large portion of their time in weaving baskets of split cedar, far exceeding any similar Indian work which we have chanced to see elsewhere, both in the coloring and the very ingenious combination of figures. Some of these baskets are so closely woven out of the dried inner bark of the willow-tree that they will hold water without leaking; the author also saw drinking-cups thus manufactured. Visitors rarely fail to bring away interesting specimens of native work in this particular line; the fine straw goods of Manila do not excel this in delicacy and beauty. In addition to this attractive basket-work from the hands of the women, the men of the tribe exhibit their natural skill by carving silver bracelets (made from dollar and half dollar coins), miniature totem-poles, horn and wooden spoons, baby rattles and canoes, in a very curious and original manner. Once a fortnight, during the summer season, on the arrival of an excursion party by steamer from the south, the natives are, as a rule, completely cleared out of their entire stock of these productions, and they do not fail to realize fair prices, enabling them to live very comfortably.

Though Sitka is the capital of the Territory, Juneau is the principal settlement and headquarters of the mining interests, containing over seven hundred white residents. We have seen no statistics of the annual rainfall here, but can well believe it to be what a certain person told us it was, namely, over nine feet. It seemed to us that the permanent residents should be web-footed. The cause of this humidity is very evident. There arises from the warm Japanese Current on the coast a constant and profuse moisture. This the winds convey bodily against the frosty sides of the neighboring mountains, and then it is precipitated as rain; at certain seasons of the year it continues for weeks together.

There is compensation even in the fact of this large annual rainfall, which at first thought seems to be such an objection to this district. The gold-bearing quartz which prevails here is treated, necessarily, by what is known as the wet process, requiring at all times an ample supply of water. One successful superintendent told the author that ore which is here so profitable would be in a dry region, like that of some portions of our Western States, worthless, or comparatively so, as it would have to be transported in bulk to a more favorable locality. It seems to require two rainy days to one pleasant one, which is about the average proportion in the year, to provide sufficient water to work these large deposits properly. The system of disintegrating, and of reclaiming the precious metal from the flint-like combination in which it is held is marvelous in detail, evincing the rapid progress which has been made in mechanical and chemical processes in our day.

It is found that June, July, and August are the favorable months for the traveler to turn his face towards the shores of Alaska, this being the season when the pleasant weather is most continuous. It is not extremes of cold, but an over-abundance of moisture in the shape of rain, which one must prepare for. An ample waterproof outside garment will be found at times very serviceable.

The Treadwell gold mine, just opposite Juneau, on Douglas Island, is undoubtedly the largest in the world, running at the present time two hundred and forty stamps, the mill and machinery having cost over half a million dollars; and though the author has visited the mines of Colorado, Montana, California, New Zealand, and Australia, he has certainly never seen its superior in capacity and golden promise. It is a true gold-bearing quartz visible at the surface, four hundred and sixty-four feet in width. The company owns three thousand running feet upon this deposit, — it can hardly be called a vein, — parts of which have been tunneled and shafted simply to test its extent, showing it to be practically inexhaustible, no bottom having been found to the gold-bearing quartz, nor any diminution in the quality of the ore. The mill is run upon this quartz the whole year, but as it is owned by a private corporation, and there is no stock for sale, the exact output of the mine is not known. The writer feels safe in saying, however, that no such body of gold-bearing quartz is known to be in existence elsewhere.

The laborers do not have to work in dark, underground channels; all is above ground, and in the season when darkness comes it is dispelled by electric lights. No timbering or shafting is required; it is simply an open quarry. Captain John Codman, after visiting the mine, writes: " We walked through the golden streets of this New

Jerusalem, with golden walls on either side, and wondered what men could do with so much money." It is not a little confusing to a stranger, when he first enters the great Treadwell Mill, to be greeted by the deafening cannonade of two hundred and forty stamps. Each stamp weighs nine hundred pounds, and the crushing capacity of the whole mill is seven hundred and twenty tons per day. The gold is shipped to the mint in San Francisco in the form of bricks worth from fifteen to eighteen thousand dollars each.

Douglas Island was named by Vancouver in honor of his friend the Bishop of Salisbury, and is eighteen miles long by about ten in width. This remarkable quartz vein is believed to run the whole length, though it is not always visible at the surface. Governor Swineford, in one of his annual reports, expresses his belief that ere long the gold produced in this section alone will exceed annually the amount which was paid to Russia for the whole of Alaska. This island, like Baranoff upon which Sitka is situated, is absolutely seamed with gold-bearing quartz, and has been carefully prospected and recorded by people interested in mining. Three hundred laborers are regularly employed at the Treadwell Mill, whose seven owners are opulent citizens of San Francisco. The work is prosecuted with great system and intelligence. The quartz of this mine is not so rich as that of many others, yielding on an average less than ten dollars to the ton, but it is so immense in quantity, and is so easily worked, that the aggregate yield of the precious metal is indeed remarkable. The mill turned out in the first twelve months after it was started seven hundred and fifty thousand dollars in bullion, and is probably producing at this writing three times that amount yearly.

The mine is admirably situated for the purpose of receiving or shipping freight, as vessels drawing twenty feet of water can lie alongside of the rocks which form the natural shore less than one hundred yards from the quartz mill. We were informed that sixteen million dollars have been offered and refused for this property. The would-be purchasers were members of a French syndicate. The agent says that the owners have but one price, namely, twenty-five million dollars, and they are in no haste to part with their property even at that sum. On the mainland, just across the channel from Douglas Island, three or four miles back of Juneau, is Silver Bow Basin, where there are gold deposits of vast extent and richness. Here quite a population is engaged in placer and quartz mining. The miners present a motley crowd with their picks, shovels, and red shirts, many with a stump tobacco pipe between their lips, and all with eager faces.

A spacious and thoroughly equipped quartz mill is being erected by a Boston company of capitalists for the purpose of developing a large property which it is thought will nearly equal the Treadwell in its output of the precious metal. This is known as the Nowell mine, and it is said that the quartz assays one hundred dollars and over to the ton. Silver Bow Basin is a small round valley lying in the lap of the mountains, accessible through a deep gulch behind the town. It is surrounded by noisy waterfalls, which supply just the needed power for manipulating the gold quartz. Across the range is another rich mineral locality, known as Dix Bow Basin.

On Admiralty Island, near the northwest end of Douglas Island, opposite Takou Inlet, there has lately been discovered several gold deposits which are owned by a Boston company. The prospectings upon some of this well-defined vein have developed a percentage of gold to the ton so large that we hesitate to specify it. "Thirty years ago," said Mr. Thomas S. Nowell to us, "the mines of Alaska would have proved comparatively valueless; the machinery and process that are now so successfully applied to reducing the ores were then unknown. The great economy and consequent profit is derived from late discoveries which are now perfected, producing machinery which works as though it had the power of thought."

The names of several other profitable mining enterprises in this vicinity might be given, but we have said enough to indicate the great mineral wealth of this portion of the Territory, and to justify our title of THE NEW ELDORADO. There are abundant gold indications all along the coast, as well as upon the islands. In the sands of any considerable stream between Cape Fox and Cook's

Inlet the "color" of gold can be obtained by the simple process of panning. The question is not where gold can be found in Alaska, for it seems to be wonderfully and abundantly distributed, but as to what localities will best pay to expend capital in developing. A number of abandoned claims show that the failure to realize a satisfactory profit in gold mining by eager, impatient, and unreasonable individual seekers without proper machinery is as frequent as in any other business enterprise awkwardly planned. This is as apparent in Africa, Australia, and California as it is in this region. The Treadwell mine on Douglas Island is in latitude 58° 16' north, just about on a line with Edinburgh, Scotland.

We quote once more Mr. Nowell's own words: "The mountains of Alaska abound in gold-bearing quartz, the extent of their deposits exceeding any similar discoveries in the world. There is without doubt more gold-bearing quartz on Douglas Island alone, which can be worked at a handsome profit, than ten thousand stamps could crush in a century; a well-defined vein from two to six hundred feet wide traversing the island for at least from six to eight miles."

There is a missionary family, supported by the Quaker persuasion, located at Douglas Island, whose earnest effort in civilizing and teaching the natives has been crowned with considerable success. The self-abnegation and conscientious labor of these people are truly worthy of all commendation.

Soon after leaving Juneau, when near the head of Lynn Channel, the grand Davidson glacier comes into view, filling the space between two lofty mountains. It measures twelve hundred feet high by some three miles in breadth, being as wide as a frozen sea and as deep as the ocean. While looking upon it one is overawed by a sense of its immensity and grandeur, as it seems hanging, poised, ready to drop into the fathomless sea. Where we pass it there intervenes a terminal moraine overgrown with trees and green foliage, which contrasts vividly with the icy background formed by the glacier. The glaciers of Europe are mere pygmies in comparison with this marvel, which is named after Professor Davidson, who has carefully explored and described it. Both the Muir and Davidson glaciers are spars of the same great icefield, which has an unbroken expanse large enough to lie over the whole republic of Switzerland. The Muir glacier will be reached presently in Glacier Bay.

Soon after leaving the Davidson glacier we are in Pyramid Harbor. This is the region of the Chilcats, who were formerly one of the most warlike tribes in the Territory, but who seem to have outlived their belligerent propensities. Their rude, but picturesque cabins dot the neighboring shore. The little settlement here consists mostly of bark huts and a substantial trader's store, together with an extensive and successful fish-cannery. The product of the latter is over a million pounds of fish per annum, the whole being engaged for 1889 to a Liverpool firm. This amount is shipped in seventy thousand cases of about fifty pounds each; the fish are packed in tins holding a pound each. This is an average amount as regards various factories on the coast, though some very much exceed it. The Indians now cheerfully accept employment from the whites, and gladly receive the regular wages which may be agreed upon. They appear to be the best carvers on the coast, and have an abundance of their handiwork to sell to the interested white visitors. These articles consist of carvings in ivory (walrus' teeth), decorated sheep-horns, copper and silver bracelets, bows, arrows, and spearheads. As engravers on copper and silver the Chilcats excel all other people of the Northwest. Some of their women wear a dozen narrow bracelets on each arm, all of home manufacture. They are also skillful in making ear-rings, and ornamental combs out of ivory and sheep's horn. As successful imitators they are remarkable, and will almost exactly reproduce any design which is given to them as a pattern. It seems strange that so aggressive and warlike a tribe should be skilled in carving and many mechanical productions.

Certain people have bestowed much honest but needless sympathy upon these "poor abused Indians." Such persons may be assured that they are amply able to look out for themselves and their own interests, as regards all material matters. No white man can get any advantage over an Alaskan native in the way of trade; they are sharpness itself in such things. For instance, these Chilcats a few years since observed that the white traders were particularly desirous of obtaining black fox skins, and that for such pelts they would willingly pay a handsome advance over skins of other colors; a fine skin of this sort bringing as high as thirty dollars, while the common red ones were not worth quarter of that sum. The innocent natives soon began to produce the black skins in large quantities and received their pay accordingly. Surprise being at last excited by the remarkable abundance of the black pelts, an explanation of the cause was sought, when it was finally discovered that by a secret process of dyeing the natives had made the red fox skins temporarily into black. This was done so cunningly that nothing but a careful examination would detect the outrageous cheat, and not anticipating anything of the kind the traders were not on their guard. Of course no dyeing process which they possessed was of a permanent nature as applied to pelts, and these black furs when they came to be prepared for market rapidly resumed their natural color. When charged with this gross deception, the Chilcats assumed the most innocent expression and denied any knowledge whatever in the premises, only saying: "Fox, him get black before him caught," thus lying concerning their trickery as volubly as any white rogue might have done.

We are told of several of these tricks played off by the "poor abused Indians," one instance of which we remember as having occurred at Fort Wrangel, illustrating the "aptitude" of the aborigines, not to give it any harder name. It seems that a kindly disposed missionary, by exercising great patience, had taught some Indians to read and write, and in the consciousness of his own intentions felt amply paid by the goodly progress of his pupils. One of these young men, not over twenty years of age, was especially curious about arithmetic, and made considerable progress in figures in a very short time. He was soon after hired by the superintendent of a fish-canning establishment as a special assistant, with good wages. Being given a note or due-bill of twenty-five dollars by his employer, he quickly saw his chance, and adroitly *raised* the figures to two hundred and fifty dollars, got the bill cashed at one of the neighboring trading establishments, and suddenly disappeared with the proceeds thereof. He has not since been seen.

The Chilcats have, until within a few years, forcibly kept the natives of the interior away from the coast and the white men, thus monopolizing the land fur-trade by acting as middle-men, so to speak, but this embargo is now entirely removed. By this and some other means, being naturally thrifty and saving, they have come to be the richest and most independent tribe of Indians in the Northwest. Their women manufacture the famous and really very fine Chilcat blankets, which are slowly woven by hand on a primitive loom. The base of these blankets is the long fleece of the mountain goats, which is tastefully manufactured and ornamented, reminding one of the domestic Oriental work offered for sale in the Turkish bazaars of Cairo. The Chilcat blankets readily bring forty dollars apiece, and the best of them are sold for double that sum. They are ordinarily about six feet long by four broad, having in addition a long, ornamental fringe at each end. The colors are black, white, yellow, and a dull blue, the coloring matter being also of native manufacture. These blankets used to be heirlooms in the aboriginal families before the cheap woolens of commerce were introduced among them, since when they have become annually more and more scarce, and are now purchased only by visitors to carry away as curiosities. Even at the highest price realized for them, if the maker's time were to be reckoned of any account, the sum is a sorry pittance for one of these blankets, which to properly finish will employ six months of a woman's time.

Pyramid Harbor, in latitude 59° 11′ north, is the most northerly point reached by the excursion steamers on this part of the coast. The place takes its name from a prominent conical formation upon an island within its borders. The cluster of houses, cabins, and the canning factory which make up what is known as Pyramid Harbor are situated upon a broad plateau on a sandy beach, at the foot of a mountain which towers three thousand feet heavenward, covered with trees to its summit and beautified by a bright, dashing waterfall visible from near the apex to the bottom. This affords both a healthful water supply for domestic use and a motor for the factory. The broad plateau, three or four miles in length and one wide, grass-grown, and covered with low shrubbery, is beautified by a floral display of great variety, including wild roses, sweet peas, columbines, white clover, and other varieties, having also an unlimited amount of berries. The wide mouth of the Chilcat River, which makes into the bay a mile from this settlement, is a swarming place for the salmon. The river is very shallow and not navigable for anything but native canoes. Twenty miles inland on its bank is a large, independent settlement of the Chilcat tribe.

On the mountain side, nearly half way up, just back of the steamboat landing at Pyramid Harbor, there is a small plateau not more than ten or fifteen feet square, entirely bare of timber, but closely surrounded by dense woods. This spot is quite inaccessible to human feet. A large cinnamon bear shows himself here often during the daytime. A clear, sparkling stream of water comes from far above this place, rushing by one corner of it, and hither comes Bruin to slake his thirst. He knows very well that he is out of the hunter's reach, and he is actually beyond rifle range. He looks at that distance skyward no bigger than a good-sized Newfoundland dog, but to appear of such proportions to us so far below he must be a very monster. Several attempts have been made by the whites to get near enough to shoot him, but without success. The bear sat upon his haunches when we saw him and peered down upon us as we stood on the deck of the Corona with a cool insolence which must have been born of a consciousness of entire safety. By using a good glass his mammoth size became more apparent, showing that even when upon his haunches with his body erect he must have measured about six feet in height.

A settlement opposite to Pyramid Harbor is known as Chilcat, where two large fish-canning establishments afford profitable occupation for quite a number of the residents, both natives and whites. New canning factories are being located in several places between Dixon Entrance and this point, the supply of salmon being absolutely unlimited; the demand only is to be considered. The quantity shipped from here annually to San Francisco for distribution is enormous, almost beyond belief, and is steadily increasing. In addition to this profitable and important industry twelve thousand barrels of salted salmon were exported last year from Alaska to southern Pacific ports. The scenery about Pyramid Harbor is arctic: the precipitous cliffs are covered with snow on their tops, and range upon range of snowy mountains frame in the bay.

CHAPTER XX.

Glacier Bay. — More Ice Bays. — Majestic Front of the Muir Glacier. — The Bombardment of the Glacier. — One of the Grandest Sights in the World. — A Moving River of Ice. — The Natives. — Abundance of Fish. — Native Cooking. — Wild Berries. — Hooniah Tribe. — Copper Mines. — An Iron Mountain. — Coal Mines.

FROM Pyramid Harbor we turn southward for a short distance, and then again towards the north, soon reaching the ice-strewn waters of Glacier Bay, an open expanse of ocean fully thirty miles long by from ten to twelve in width. This locality is thus named because of the number of glaciers which descend into it from the southern verge of the frozen region. The still surface of the water reflects the Alpine scenery like burnished silver, only ruffled now and again by the icebergs launched from the majestic front of the Muir glacier, which fall with an explosion like the blasting of rocks in a stone quarry. It is curious to watch these enormous masses of ice rise to the surface after their first deep plunge, see them settle and rise again until their equilibrium becomes fixed, and then slowly float away with their imperial colors displayed, to join the fleet gone before. They seem to exhibit in their vivid colors a radiant joy at release from long imprisonment. It was a gloriously bright day on which we approached the

Muir glacier, the sun pouring down its wealth of light and warmth to temper the crisp morning air. A side-wheel steamer could not have made headway among the hundreds of floating icebergs; but the Corona wound in and out among them in safety, piloted by Captain Carroll's skillful direction, occasionally leaving the color of her painted hull along their sides by chafing them.

The ship was brought within fifty rods of the glacier's threatening front, which was about three hundred feet in height above the water, standing like a frozen Niagara, and the lead showed it to extend four hundred feet below the surface, making an aggregate of seven hundred feet from top to bottom. What a mighty power was hidden behind the dazzling drapery of its iridescent façade!

Standing upon its surface a short way inland, one could hear from its depths what seemed like shrieks and groans of maddened spirits torturing each other, as the huge mass was crowded more and more compactly between the two abutting mountains of rock through which it found its outlet. The roar of artillery upon a battlefield could hardly be more deafening or incessant than were the thrilling reports caused by the falling of vast masses of ice from the glacier's front. Nothing could be grander or more impressive than this steady bombardment from the ice mountain in its resistless progress towards the sea. Neither Norway nor Switzerland have any glacial or arctic scenery that can approach this bay in its frigid splendor. No natives are to be seen; not a sound falls upon the ear save the hoarse cannonading of the glacier. The white, ghostly hue of the surroundings are startling; even the daylight assumes a certain weird, bluish tint, heightened by shimmering reflections from the ice-chasms and crevices.

The author, in a varied experience of many parts of the world, recalls but two other occasions which affected him so powerfully as this first visit to Glacier Bay in Alaska, namely: witnessing the sun rise over the vast Himalayan range, the rooftree of the globe, at Darjeeling, in northern India, and the view of the midnight sun from the North Cape in Norway, as it hung over the Polar Sea. Our power of appreciation is limitless, though that of description is circumscribed. Here both are challenged to their utmost capacity. Words are insufficient; pen and pencil inadequate to convey the grandeur and fascination of the scene.

Lieutenant Frederick Schwatka tells us that a veteran traveler said to him as they stood together on the ship's deck regarding the scenery in this remarkable bay: "You can take just what you see here and put it down on Switzerland, and it will hide all there is of mountain scenery in Europe. I have been all over the world, but you are now looking at a scene that has not its parallel elsewhere on the globe." The estimate has been made by experienced persons that five thousand living glaciers, of greater or less dimensions, are now steadily traveling down towards the sea in this vast Territory of Alaska.

Glacier Bay is always full of vagrant icebergs which are of blinding whiteness when under the glare of the midday sun. The variety of colors emitted by the bergs is charming to the eye, the prevailing hues being crystal-white mingled with azure blue, a faint touch of pink appearing here and there, together with dainty gleams of orange-yellow. Where a large smooth surface is presented, the prismatic shimmering is like that of starlight upon the water. The variety in the shape of the bergs is infinite. Some of them exhibit singularly correct architectural lines, some resemble ruins of ancient castles on the Rhine, others, with a little help of the imagination, represent wild animals in various attitudes, or hideous Chinese idols with open mouths and lolling tongues. Sea birds hover over and light in large numbers upon the opalescent masses. Ranging alongside of a tall berg, a fall and tackle was rigged out from the yard-arm of our steamer, while men were sent to cut large blocks of ice from the hill of frozen water. Two weighing nearly a ton each were hoisted on board to keep our larder cool and fill the ship's ice-chest. The ice was pure as crystal, and fresh as a mountain stream.

"Why don't you go nearer to the glacier?" asked one of the passengers of the captain.

"Because I think we are quite near enough," was the quiet reply.

"Those avalanches don't reach more than thirty

or forty feet from the face of the ice cliff," continued the passenger.

"True," was the reply, "but they do not conconstitute the only discharges from the glacier."

"Why, where else can they occur but from the face," asked the inquirer.

"Shall I tell you a certain experience which I had near this very spot?" asked the captain.

"What was it?" inquired a dozen eager voices.

And then the captain told the group of listeners that when the Corona was here last season, laying just off the Muir glacier, those on board were startled by the sudden appearance of a huge mass of dark crystal, as large as the steamer itself, which shot up from the depths and tossed the ship as though it had been an egg-shell. Passengers were thrown hither and thither, and some were severely bruised. It was a berg broken off from the bottom of the ice mountain, four hundred feet below the surface of the water. Had it struck the ship in its upward passage, immediate destruction must have followed, and the steamer would have sunk as quickly as though she had been blown up with gunpowder.

Mount Crillon, Mount La Perouse, and Mount Fairweather are all visible from Glacier Bay, the latter rising in the northwest so high above the intervening hills that all its snowy pinnacles are clearly defined.

The great glacier which forms the prominent feature of this bay was named after Professor Muir, state geologist of California. It has a front three miles wide, and has been explored to a distance of forty miles inland. The top surface is tossed and broken by broad fissures so as to be impassable, unless one goes back at least a mile from its toppling and dangerous front. This glacier exceeds anything of the sort this side of the polar zone, and is fed by fifteen other glaciers, so far as it has been explored, towards its source among the lofty snow-fields. In walking upon its surface great care should be observed. A thin crust of snow and half-melted ice is often formed over fissures into which one may easily be precipitated. One of the party from the Corona, a lady, was thus engulfed for a moment, escaping, however, with a thorough wetting and some slight bruises, together with a very large measure of fright. This lady was temporarily in charge of the pilot of the steamer, hence it was very generally remarked that he was doubtless a good ship's pilot, but a poor one for navigating glaciers.

From carefully conducted measurements it is known that this immense body — frost-bound, transparent, and resistless — is moving into the sea, during the summer months, at the rate of forty feet in every twenty-four hours, and discharging in that time one hundred and forty million cubic feet of ice into the bay. It is not necessary for us to discuss the cause of this regular, uniform movement of the enormous mass; it may be brought about by either dilation or gravitation, both of which are most likely active agents to this end, but certain it is that the glacier moves forward as described.

One could have passed days in studying the grandeur and beauty of the Muir glacier, in watching its slow but steady advance, its tremendous avalanches, its rolling, thunder-like discharges, its irregular, translucent front decked with amethyst and opal hues by the afternoon sunlight, but time was to be considered, the day was closing, and we finally steamed reluctantly away. Even after we had lost sight of the great frozen river, we heard its evening guns echoing among the mountains, faint and fitful from the growing distance.

We pause for a moment, thoughtfully, to recall the brief hours passed in that boreal atmosphere, crowded to repletion with wonderful experiences, where the ice deposited during the glacial period is slowly wasting and wearing away, exposing giant cedars which have been buried for ages upon ages, a revelation and a process which we may nowhere else behold. There is no touch of civilization here; the quiet and solitude is unbroken, save by the thunder of the bergs breaking their long imprisonment. Somehow one feels older, grayer, sadder, after witnessing these great and startling throes of Nature, phenomena which have been in operation thousands of years. It reminds the observer only too forcibly how infinitesimal is the space he occupies upon this planet, and how utterly insignificant is his personality in the vast scheme of the universe. Travel, while teaching us numberless grand and beautiful truths, solving many mysteries and vastly enlarging our mental grasp, does not fail also to impress upon the most conceited the important and priceless lesson of humility. But let us banish brooding thoughts, and be glad for a little space; to-morrow the night cometh!

Among the evidences of the slow but steady receding of the glacier we have Vancouver's record that he was unable to enter this bay in 1793, which is now navigable for over twelve miles inland. Once the ice field was level with the mountain tops, now it has melted until the peaks are far above its surface. Professor Muir tells us that in the earlier days of the ice-age this glacier stood at a height of from three to four thousand feet above its present level! Centuries hence the place of the glacier will doubtless be occupied by a flowing river, and the land will have entirely thrown aside its mantle of ice and snow. What a revelation this bay would have been to Agassiz! After an arduous half day's climb, from the summit of the Muir glacier nearly thirty others are to be seen in various directions, all steadily forcing their resistless way towards the sea, slowly consummating the purpose of their existence. How far glacial action has been concerned in determining the topographical conditions of the globe will long be, as it has long been, a subject for deep scientific study.

At first thought it seems impossible that a substance like ice, often brittle as glass and as inelastic as granite, can move as though it were fluid. The motion of the giant mass is doubtless facilitated by subglacial streams issuing from its bottom into the bay. The water flowing from two sources of this character manifests itself at the surface on each corner of the ice-front, where it comes bubbling up with great force from the bottom, a distance of from sixty to eighty fathoms. As we lay in front of the grand façade what a revelry of color was spread before us! The immense and towering wall of ice seemed to throb with the softening rays of the sun, penetrating each broad fissure and narrow rift, all luminous with blue and gold.

Scidmore Island was pointed out to us, a green hilly land, near the mouth of the bay, named after Mrs. E. R. Scidmore, who has written so admirably about Alaska. Another island was designated whereon a silver mine of great promise has lately been successfully located and tested, yielding results surpassing the most sanguine anticipations of the owners.

All through this region one is constantly impressed with a sense of vastness, everything seems so stupenduous; Nature is cast in a larger mould than she is in other sections of the world. The islands strike one as continental in dimensions, the rivers are among the largest on the globe, the ocean channels are the deepest, the primeval forests are made up of giant trees and cover thousands of square miles, the mountains are colossal, and the glaciers are elsewhere unequaled. It is a land of wonders, strange, fascinating, and beautiful.

The natives of this latitude are robust and

hearty in appearance, their regular food supply being such as to sustain them in a good physical condition. Seal and fish oil are cheap and abundant, and enter into all of their cooking combinations. During the ripening season the wild berries, which are remarkably abundant, are gathered by the bushel, giving employment to the youthful portion of the community. Large quantities are dried for winter use, but during the bearing season the people almost live upon them, always adding a portion of oil as a condiment. Game, such as deer, bears, mountain goats, and wild geese, is very plenty a little way inland. These are hunted and supplied to the whites by the aborigines, but they do not themselves seem to care particularly for meat of any sort so long as they can obtain plenty of fish and oil. At Sitka and Fort Wrangel fine large codfish are retailed at five cents each, a twenty pound salmon costs in the season ten to fifteen cents, and halibut sell at about the same rate according to size. These latter average from eighty to a hundred pounds in weight on this coast, and in some parts of the waters bordering western Alaska they are twice that size. Ducks are to be had at ten and fifteen cents per pair, wild geese at fifteen cents each, and so on. The natives are preëminently fish-eaters, and are as a rule well developed about the chest and shoulders, though the lower parts of their bodies are diminutive owing to their exercise being taken almost altogether at the paddle while sitting in their boats. The physical contrast between them and our Western Indians, who are meat-eaters, is very decided. The one lives in a canoe a large portion of his time, the other upon horseback or engaged upon long foot-marches; the one is lithe and sinewy, the other is greasy and flabby. Though the physical condition of our Western Indians is unquestionably much superior to that of the native Alaskans, yet the latter are the most intelligent.

The halibut, to which reference has just been made, is found in great abundance upon the coast at nearly all seasons of the year, and forms a large portion of the food supply of the native population, both for summer and winter. They prefer to catch these fish by means of their own awkward wooden hooks, rather than to use the steel barbed instrument of the whites. They go out for the purpose in their boats, exposing themselves in nearly all sorts of weather, anchoring upon well-known fishing grounds by making use of a stone fastened to a cedar-bark rope of their own manufacture. Having filled their canoe, which they can do in a very short time, they leisurely return to the shore, where the fish are turned over to the care of the women, who soon clean them, also removing the large bones, head, fins, and tails, after which they cut the bodies into broad thin slices, and doing so much of this business they become very expert. These slices of the halibut are hung on wooden frames, where they rapidly dry in the wind and sun, no salt being used in the process; indeed, the natives seem to have no use for salt so far as their own food is concerned, and do not eat it as a seasoning. After the halibut is thus cured, the pieces are packed away in the large cedar box which forms each family's storehouse for such food, and when wanted it is always ready, requiring but little further treatment to make it palatable to native Alaskan taste. As thus preserved the fish will now and again become putrid. This, however, is not considered by the people to detract in any degree from its excellence and usefulness, but rather to add zest to the flavor, just as a highly civilized gourmand requires his birds to be kept until they become a little "gamey" before he considers them fit to serve to himself or his guests. At certain seasons of the year the salmon are eagerly sought and eaten, both fresh and dried, but as intimated the halibut is a fish which can be caught at nearly any time, and is therefore perhaps more used than any other. There are periods when these fish also leave the coast for a short season, and against this absence the native provides as we have described. The kind of salmon which is mostly canned and prepared for export in barrels from Alaska is of a pink species, which is chosen, not because it possesses any peculiar excellence of flavor, but because the color is generally thought to be more desirable. They are not considered here, either by the whites or the natives, to be of quite so good quality as some others which abound in this region, but it is the pink salmon which the fanciful public demand, and pink salmon which they get.

All the cooking these natives seem to know anything about is to boil or stew such food as they do not consume nearly raw. Iron kettles have been in their possession for many generations, and were originally procured from the Russians. The condiment which they most affect has already been referred to, being nothing more nor less than rancid fish or seal oil, cooled and hardened into a sort of oleomargarine, the bare smell of which is sickening to the nostrils of a white person. This grease is spread liberally upon all their food and eaten with manifest relish. The inner bark of the spruce and hemlock trees is collected by the women in considerable quantities at certain seasons of the year, and is eaten by them, both in the green and dried state, after being dipped in this grease as described. The Sitka Indians make a most atrocious salad of seaweed mixed with seal-oil, sometimes adding the roe of herring, of which peculiar mixture they partake with ravenous appetites, the roe having been purposely kept until it is nearly or quite putrid. The salmon-berry, while it is in season, is a most welcome and wholesome addition to their rather circumscribed larder. This berry is a sort of cross between a strawberry and a blackberry, though it is larger than the average of these delicious berries as they grow in the woods of New England. Hundreds of barrels of the native cranberry are gathered by the aborigines and shipped annually from here to San Francisco; they are smaller than the cultivated berry bearing the same name, which is grown in our Eastern States. The wild strawberries found among these islands and on the mainland excel in flavor the highly cultivated berry of our thickly-settled States, and may be found growing in abundance in the very shadow of the glaciers.

The natives hereabouts have no domestic animals except a multitude of dogs of a mongrel breed; wolfish-looking creatures; which are of no possible use, dozing all day and howling all night. At the north the regularly bred Eskimo dog is a very different animal, quite indispensable to his master, and invaluable in connection with sledge traveling.

The tribe occupying the region near to Glacier Bay is known as the Hooniahs, an ingenious and industrious people, who manufacture bracelets, spoons, and various ornaments out of silver and copper. Some of the men of this tribe wear a ring in their noses, like the women, but this seems to be going slowly out of fashion. We were told that the men have as many wives as they choose to take, and that they are not always careful to properly discriminate between other men's and their own, an act of dereliction from propriety which is, however, by no means confined to savage life. A great laxity in morals is also said to prevail among most of the tribes from Behring Strait southward to the Aleutian group of islands. Let us not, however, be too censorious in judging them; if their virtues are found to be in the minority, is not this also the case with most communities which boast the elevating advantages of culture and civilization?

It has been known for a century more or less that masses of pure copper were found by the aborigines along the course of Copper River, which flows into the Pacific Ocean midway between Mount St. Elias and the peninsula of Kenai. The natives exhibited one mass of pure copper, as naturally deposited, weighing over sixty pounds. The character of this mineral closely resembles that of our Lake Superior district, and there is every indication of its abundance in this region, not alone on Copper River, but in several districts and islands. The natives have utilized the article for many generations in the manufacture of personal ornaments, and for making various useful household utensils, such as stewpans and small kettles. Any permanent rise in the market value of copper would stimulate the development of the copper mines of Alaska to compete with other portions of our country. Petroleum is also found on Copper River, forcing itself to the surface from some underground reservoir, and again near the Bay of Katmai. This product was largely used by the Russians for lubricating purposes.

Professor Davidson discovered in this vicinity an iron mountain some two thousand feet high, which was so full of magnetic ore as to seriously affect his calculations and derange his compass. Mr. Seward said of the same vicinity: "I found there not a single iron mountain, but a whole range of hills the very dust of which adhered to the magnet." There is plenty of coal also, and with these two articles in juxtaposition a great industry may ultimately be the outgrowth. Viewed as a sure foundation of commercial and manufacturing prosperity, coal and iron will prove, in the long run, to be worth nearly as much to Alaska as her abundant and inexhaustible gold supply.

Captain J. W. White of the United States revenue marine says: "I have seen coal veins over an area of forty or fifty square miles so thick that it seemed to me to be one vast bed. It is of an excellent steam-producing quality, having a clear white ash. The quantity seemed to be unlimited. This bed lies northwest of Sitka, up Cook's Inlet which broadens into a sea in some places." Nature has provided fuel in limitless quantities for this great Territory, both in the form of coal and of wood, each of which is of the most available character, both as regards the quality and the convenience of location.

In speaking of the rich and varied prospects of the country, let us not forget to mention the abundance of pure white, statuary marble, which exists here in immense quarries, near the site of which there are numerous safe and commodious harbors, with great depth of water, inviting the commerce of the world. We need not send to Italy for a fine article in this line; the choicest product for statuary purposes is here upon our own soil. While these sheets are going through the press, the fact that a valuable quicksilver mine, which was discovered at Kuskoquin some years ago, now proves to be of high grade and purity, is published to the world at large. If so, this is extremely providential, as there is now a constant demand for mercury in the treatment of the gold-bearing quartz of the numerous mines hereabouts.

The studied effort of certain writers to depreciate the value of the Territory of Alaska in nearly every possible respect seems very singular to us, and is altogether too obvious to carry conviction with it. The great amount of gold now being realized every month of the year, the millions of cured salmon and cod annually exported to other sections, together with the rich furs regularly shipped from the Territory, counted by hundreds of thousands, must cause such people a degree of mortification. One of these writers put himself on record by saying not long since that gold did not exist in the Territory in paying quantities. Yet there is a standing offer of sixteen million dollars for the Treadwell gold mine on Douglas Island, while within eight or ten miles of it, at Silver Bow Basin, on the mainland, is another gold mine, as has been shown, owned and worked by a Boston company, nearly as valuable.

Referring to this auriferous deposit on Douglas Island, Governor Swineford says, in his official report to the government for the year 1887: "It is without doubt the largest body of gold-bearing quartz ever developed in this or any other country."

At last we prepare to turn our backs upon the

home of the glaciers and the locality of the most remarkable gold deposits of the Northwest, surfeited with wonders, and actually longing for the sight of something intensely common, satisfied that the tourist who makes the voyage from Tacoma to Glacier Bay through the inland sea has the opportunity of beholding some of the grandest scenery and natural phenomena on the globe.

CHAPTER XXI.

Sailing Southward. — Sitka, Capital of Alaska. — Transfer of the Territory from Russia to America. — Site of the City. — The Old Castle. — Russian Habits. — A Haunted Chamber. — Russian Elegance and Hospitality. — The Old Greek Church. — Rainfall at Sitka. — The Japanese Current. — Abundance of Food. — Plenty of Vegetables. — A Fine Harbor.

FROM Glacier Bay our serpentine course lies southward through the countless sounds, gulfs, and islands of various shapes and sizes to Sitka, the New Archangel of the Russians, Sitka being the aboriginal name of the bay on which the town is situated. This is the most northerly commercial port on the Pacific coast, and lies at the base of Mount Vestova on the west side of Baranoff Island. The island is eighty-five miles long by twenty broad, situated thirteen hundred miles north of San Francisco.

On the 18th of October, in the year 1867, three United States men-of-war lay in the harbor, namely, the Ossipee, the Jamestown, and the Resaca. It was a memorable occasion, for on that day the Muscovite flag was formally hauled down and the Stars and Stripes were run up on the flagstaff of the castle amid a salvo of guns from the ships of both nations, thus completing the official transfer of the great Territory of Alaska from Russian to American possession.

Up to this time the government of the country had been virtually under the control of the rich fur company chartered by the Tzar. Any policy at variance with its purposes was treason; immigration, except for its employees, was rigorously discouraged; the imperial governor was actually salaried by this great monopoly, while his public acts were subject to its approval or otherwise. With the date above given this condition of affairs ceased and a new régime began. Though no radical change immediately took place, still the atmosphere of our Union gradually permeated these regions, our flag freely floated everywhere, and our few officials assumed their responsibilities, administering the laws of the Republic mercifully as regarded the natives, but still with that degree of firmness which is imperative in dealing with a half-civilized race.

One cannot but conjecture what must have been the secret thoughts of the thousands of aborigines on this occasion, as they witnessed the ceremony of transferring Alaska from their former to their new masters. It was an event of immense interest, of most vital import to them, but yet one in which they were entirely ignored. They knew the significance of that change of flags, of that roar of artillery, emphasized by other naval and military movements, but they had no voice whatever in the agreement by which they were virtually bought and sold like so many head of cattle, and their native land bartered for gold. We leave the reader to moralize over this aspect of the matter, a fruitful theme for the political economist. With this change of government came a new people; the majority of the Russians promptly left the country, and their places were taken by Americans.

Sitka, the capital of the Territory, is sheltered by a snow-crowned mountain range on one side, and protected from the broad expanse of the Pacific on the other by a group of many thickly wooded islands. The waters of the harbor are as clear as a mountain stream, so that, as in sailing over the Bahama Banks, one can see the bottom many fathoms down with perfect distinctness, where the myriad curiosities of submarine life attract the eye by their novel and varied display. Among other tropical growth, sponges, coral branches, and long rope-like algæ are seen, planted here doubtless by the equatorial current which so constantly laves these shores. The town lies clustered near the shore, forming a pleasing picture as one approaches from the sea. The most prominent feature is the castle, not a battlemented, ivy-covered, mediæval structure, but a severely plain, weather-beaten, moss-grown, dilapidated affair, which crowns a rocky elevation of the town. It is a hundred and forty feet long by seventy deep, constructed of huge cedar logs which are securely riveted to the rock by numerous clamps and bolts. This was for many years the grand residence of the Russian governors, — after the capital was removed from St. Paul, on the island of Kodiak, — several of whom were of the Muscovite nobility and brought hither their wives and daughters to live with them in this isolated spot. One can hardly conceive of a greater social contrast than naturally existed between St. Petersburg and this half savage hamlet of Baranoff Island. For delicate and refined ladies, such a change from court life must have been little less of a hardship than actual banishment to dreaded Siberia.

It is not surprising that resort was had to rather desperate means whereby to beguile the weary hours. Many fell victims to gambling and strong drink. The Russians, under nearly any circumstances, fail to be good examples of temperance, and here cognac and vodhka flowed free as water. To some of their official feasts and celebrations the native chiefs were invited, and terribly demoralized by the potency of the viands to which they were totally unaccustomed. Nor can it be wondered at that, being occasionally supplied with this fire-water, the natives now and again broke out in open revolt, which ended more or less seriously both to the Russians and themselves. It will be remembered that once during the early times the natives rose in a body and massacred or drove every foreigner off the island, an act of savage patriotism which cost them dearly.

Every "castle" must have at least one haunted chamber, and we are told that this of Sitka was no exception to the general rule. The story concerning the same is variously told by different persons, but we will give only the version we heard. It seems that half a century and more since, the

Russian governor's family included a beautiful and accomplished daughter named Eruzoff, who was, at the time the event occurred which we are about to relate, but twenty years of age. There were on her father's official staff two young noblemen of St. Petersburg, Nicholas and Michael Burdoff, about twenty-five years of age respectively. They were cousins, and had been ardent and intimate friends from childhood. Both of the cousins fell deeply in love with the governor's daughter, who, in her delicacy, showed no preference between them. The young men grew desperate in their feelings. Never before had they disagreed about the simplest matter; it was their delight to yield to each other; but now their love for the beautiful Eruzoff made them open rivals. One day they went into the neighboring forest together, as they said, to hunt, and were absent for two days. On the evening of the second day Michael returned unaccompanied by his cousin, whom he said he had lost in the forest. He retired at once to his own room in the castle, where he was found dead in bed on the following morning, without a wound or any sign to explain the cause, though the post surgeon pronounced it to be a case of heart disease. A few days afterwards, by means of his favorite dog, the body of Nicholas was discovered in the forest with a bullet through his brain. The actual truth regarding the death of the cousins cannot be known on earth, but the chamber where Michael Burdoff breathed his last is said to be often disturbed by a ghostly visitor at midnight. Eruzoff was forced by her father to marry an official of his choice, though she was broken-hearted at the loss of Michael Burdoff, who proved to have been the one whom she loved best. She died in her bridal year.

Interesting stories are told of the grand hospitality — characteristic of the Russians — which was so liberally dispensed within this castle, in entertaining celebrated voyagers of various countries, and especially those of the United States. It has always been the policy of the Tzars to cultivate kindly feelings with our government, and Russia is still our constant friend. The upper part of the old castle was arranged for theatrical representations, while in the other apartments the nights were rendered merry with cards, dancing, and music. Rich furniture, valuable paintings, and costly plate had been brought all the way from Russia to equip this grand household among a savage race. The toilets of the ladies were perhaps a twelvemonth behind those of St. Petersburg, but their diamonds and laces were never out of fashion. Elegant chandeliers were left by these former masters of the castle, which show what the rest of the furniture must have been to have harmonized with such gorgeous ornaments. The visitor is shown the apartment occupied by the venerable Lady Franklin at eighty years of age, who came hither in search for her lost husband, the Arctic explorer.

The quaint old Greek Church with the sharp peak of Vestova as a background is a prominent and interesting edifice. Its emerald-green dome and Byzantine spire, after the home fashion of the Russians, together with its elaborately embellished interior and its ancient chime of bells, strongly individualize the structure. Some pictures of more than ordinary merit are to be seen within its walls. One representing the Madonna and Child is pronounced to be very valuable. It is kept in perfect condition by the government of St. Petersburg, which is the sole owner of all the churches of the empire, at home and abroad. The Tzar expends more money for church and missionary purposes in Alaska to-day than all the Christian sects of our country combined. For the three churches in Sitka, Kodiak, and Unalaska the sum of fifty thousand dollars annually is set aside and appropriated. Nevertheless, we believe the Training School at Sitka exercises a much higher civilizing influence, where the simplest Christian principles are taught, combined with common school studies, and where instruction is given in the daily industries of life. All concede that education and general intelligence are the mainsprings of our system of government, and that the perpetuity of its institutions depends thereon. In view of these indisputable facts let our rulers at Washington bestow liberally from out the plethoric national treasury for educational purposes in Alaska.

Most of the houses of Sitka are heavy log dwellings, some of which are clapboarded outside and smoothly finished within. In the winter season about a thousand Indians live here, the white population being composed of the usual government officials and agents, with a few storekeepers engaged in the fur traffic and general trade with the aborigines. Four or five hundred miners and prospectors gather here also in the winter, when it becomes too cold to prosecute their calling far inland, where the thermometer often falls to 20° below zero. Even this occasional extreme could be easily endured, and the work be little retarded, were suitable quarters provided. In midwinter daylight continues at Sitka for only six hours in the twenty-four, though by the first of June there is virtually no night at all; the stars take a vacation, while the evening and the morning twilight merge into day.

The author had thought, heretofore, that the rainfall at Bergen, on the coast of Norway, exceeded that of any other spot he had visited, but here at Sitka " the rain, it raineth every day." We have seen it rain harder in the tropics, but not often. The brief downpour, however, is so quickly followed by a flood of delicious sunshine that the contrast is a charming revelation. Still another effect is observable that, as rainy as it is, at certain seasons the atmosphere is still peculiarly dry. The writer was told that clothes would quickly dry under a shed during the heaviest rains. The fair weather is most likely to occur during the excursion season, so that the stranger is not apt to meet much annoyance in this respect while at the capital. The annual rainfall is recorded as being ninety inches upon this island, a degree of humidity which is attributed to the heated waters of the equatorial regions, which warm the whole coast-line of southern Alaska, insuring the mild winters it enjoys.

Scientists tell us that the effect of this warm current is equivalent to twenty degrees of latitude, that is to say, the same products which are found in latitude 40° north on the Atlantic coast thrive in this region at 60° north, which is a little higher than the latitude of Sitka. This beneficent stream, arising off the coast of southern California, crosses the Pacific south of the Sandwich Islands, and on the coast of Asia turns northward in a grand sweep, striking the shores of America, and returning finally to its starting-point. "It is this," says H. H. Bancroft, in his "History of the Pacific States," "that clothes temperate isles in tropical verdure, makes the silkworm flourish far north of its rightful home, and sends joy to the heart of the hyperborean, even to him upon the Strait of Behring, and almost to the Arctic Sea."

The abundant moisture causes all vegetation to grow most luxuriantly. "The enemies of this region, some of whom," said an official to us, "have been paid for sinister purposes to write it down, declare that it cannot be made to support a population, as vegetables will not grow here, but vegetables have been successfully grown all about us for more than fifty years." There are a plenty of domestic cattle at Sitka, where we partook of as sweet and rich milk as can be produced on our choice dairy farms at the East. The southern portions of the Territory, both the islands and the mainland, are better adapted to support a civilized white population than are the larger portions of Norway and Sweden. It may be doubted if there is anything finer in color than the June greenery of Sitka. Our first day at this unique capital had been varied by alternate rain and sunshine, but the closing hours of the day were clear and beautiful, emphasized by such a grand and brilliant sunset as is rarely excelled, the afterglow and mellow twilight lasting until nearly midnight, causing the turban of snow upon the head of Mount Edgecombe to look like Etruscan gold.

John G. Brady, United States commissioner at Sitka, writes from there as follows: " Though Alaska is no agricultural country, yet there is plenty of land for growing vegetables for a vast population which can be easily cleared and cultivated. The food of this coast is assured unless the Pacific current changes and rain ceases. Perhaps there is not another spot on the globe where the same number of people do so little manual labor and are so well fed as in Sitka." The capacity of the island to produce a large variety of garden vegetables, and of good quality, is abundantly demonstrated by a resident who gains a successful livelihood through the sale of these products grown on his own land.

The bay is very lovely and naturally recalls that of Naples, with its neighboring Vestova and its beautiful islands. Though Mount Edgecombe with its great truncated cone, situated fifteen miles away upon Kruzoff Island, is not now in active condition, a century ago, more or less, it poured forth lava, fire, and smoke enough to rival the Italian volcano which buried Pompeii in its fatal débris nearly two thousand years ago. We were told that smoke and sulphurous vapor occasionally issue from the old crater of Edgecombe, but saw no distinct evidence of the fact. As we looked at the sleeping giant we wondered if it will one day awake in its Plutonic power. The bay is said to contain over one hundred islands, which are mostly covered with a noble growth of trees, rendered picturesque and lovely by green sloping banks and shores fringed with golden-russet seaweed, bearing long, banana-like leaves. Many of these islands are occupied, some by whites, some by Indians. Japan Island, so-called, is the largest in the bay, and is situated just opposite the town. It was once improved by the Russians as an observatory, and now contains some fine gardens cultivated both by whites and natives, from whence the citizens obtain their supply of fresh vegetables. Baranoff Island itself is mountainous and thickly wooded, though there are large arable spots distributed here and there near to Sitka, dotted with wild flowers in white and gold, — Flora's favorite colors in this latitude. Never, save in equatorial regions, has the author seen vegetation more luxuriant than it is in its native condition in these islands of southern Alaska.

CHAPTER XXII.

Contrast between American and Russian Sitka. — A Practical Missionary. — The Sitka Industrial School. — Gold Mines on the Island. — Environs of the Town. — Future Prosperity of the Country. — Hot Springs. — Native Religious Ideas. — A Natural Taste for Music. — A Native Brass Band. — Final View of the Capital.

THE Sitka of to-day contains about two thousand inhabitants, but is a very different place from that which the Russians made of it. The subjects of the Tzar carried on shipbuilding, manufactured wooden and iron ware, erected an iron furnace and smelted native ore, made steel knives and agricultural tools, axes, hatches, and carpenters' tools generally. They established a bell foundry here at which many bells and chimes were cast, and shipped the products all along the Pacific coast, especially to Mexico. The Greek Church was kept up to the highest standard as regarded the national forms, and employed nearly a score of priests, which, together with some forty or fifty civil officers attached to the governor's household staff, made a considerable community of white citizens, which was a constant scene of business activity. The capital has, in some respects at least, been greatly improved since it came into our possession, but it bears unmistakable evidences of antiquity. It has been made neat and clean, which was certainly not a characteristic under its former management, the streets have been regularly laid out, and good sidewalks have taken the place of muddy pathways, while some well-constructed roads leading through the neighborhood have been perfected. Though there is not seemingly so much of local business going on as there used to be, still it is a far more wholesome and pleasant place to live in than it was in the days of Muscovite possession. In Mrs. E. S. Willard's published letters from Alaska we learn how an officer of our navy, namely, Captain Henry Glass of the United States steamer Jamestown, in 1881, proved to be the right sort of missionary to send on special duty to Sitka.

"His first move," says this lady, "was to abolish hoochinoo. He made it a crime to sell, buy, or drink it, or any intoxicating drinks. He prevailed upon the traders to sell no molasses to the Indians in quantities, so that they could not make this drink. He issued orders in regard to clearing up the native ranches, which were filthy in the extreme, and had been the scene of nightly horrors of almost every description. He appointed a police force from the Indians themselves, dressed them in navy cloth with 'Jamestown' in gilt letters on their caps, and a silver star on their breasts. He made education compulsory. The houses were all numbered and the children of each house, each child being given a little round tin plate on which was marked his number and the number of his house. These

plates were worn on a string about the neck. As
the children arrived in school they were registered.
Whoever failed to send their children were
fined one blanket. As soon as they discovered
that the captain was in earnest they submitted,
and I believe no blanket was forfeited after the
first week. The ranches have been cleaned, whitewashed,
and drained, and all is peaceful and quiet
where a few months ago it was a place of strife."

The Sitka Industrial School — or as it is better
known here, the Jackson Institution — is the most
interesting feature of the town, because one cannot
fail to realize how much good it is accomplishing
in the way of practical civilization and real
education among the natives. At this writing
there are nearly one hundred boys, and about
sixty girls and young women, who are under the
parental care of the Institution. The teaching
force consists of a dozen earnest workers, mostly
ladies from the Eastern States. Besides the ordinary
English branches taught in the school, the
girls are trained to cook, wash, iron, sew, knit,
and to make their own clothes. The boys are
taught carpentry, house-building, cabinet-making,
blacksmithing, boat-building, shoemaking,
and other industries. The work of the school is
so arranged that each boy and girl attends school
half a day, and works half a day. The results
thus brought about are admirable. The "Mission,"
as the cluster of buildings forming the
school, the hospital, the residence for teachers, cottages,
and workshops is called, is situated beside

the road leading to Indian River, overlooking the bay, the islands, and the sea, with grand mountain views on three sides. Fifteen different tribes are represented in this Sitka Industrial School. English-speaking young natives who have been trained here readily obtain good wages at the mines, in the fish-canneries, and wherever they apply for employment among the white residents of the Territory, while their influence with their tribes is very great. That the Alaskans are teachable and capable of attaining a higher and better plane of life has been abundantly proven by the successful mission of this school during the few years of its existence.

There is a small monthly newspaper published at Sitka in the interest of the Training School called "The North Star." It is inexpensively produced, and is calculated to disseminate information in behalf of the excellent mission, as well as to add interest to its local affairs. The type-setting and all the work on this little paper is done by native boys. In his last published report Dr. Sheldon Jackson says in relation to the Alaskan natives: "Christianize them, give them a fair school education and the means of earning a living, and they are safe; but without this the race is doomed. We believe in the gospel of habitual industry for the adults, and of industrial training for the children. By these means they can be reclaimed from improvident habits, and transformed into ambitious and self-helpful citizens."

The Industrial Training School at Sitka was established as a day school by the Presbyterian Board of Home Missions in 1880, with Miss Olinda A. Austin as teacher. The following fall circumstances led to the opening of a boarding department. Since then the institution has grown until there are connected with it two large buildings (one for boys and the other for girls), an industrial building sheltering the carpenter and boot and shoe shops, the printing-office and boat house, a small blacksmith shop, a steam laundry, a bakery, a hospital, and six small model cottages. Every building has been constructed by the pupils themselves under the direction of the one carpenter, who acted as their instructor. Even the domestic furniture, such as beds, chairs, bureaus, and the like, is the handiwork of these native boys. We can testify from personal observation that all is wonderfully well done, and of excellent patterns.

There is a valuable gold mine situated six or eight miles southeast of Sitka, eight hundred feet above the sea level and about a mile from deep water, on Silver Bay, where the largest ships may lie beside the shore, the wharfage having been prepared by Nature's own hand. The quartz rock is here represented to be of excellent quality, showing thirty dollars to the average ton, and there is never-failing water near at hand sufficient for running a hundred stamp-mill. Gold has been mined at Silver Bay in a primitive way for several years. Numerous other mines have been located and opened on Baranoff Island which give great promise, but this just mentioned has accomplished thus far the best results. We took notes of eleven mines upon which much work had been done, shafts sunken, and tunnels run. "The island is besprinkled with these gold-quartz veins," said an intelligent citizen to us. "Prospectors and miners have been attracted elsewhere in the Territory by still more promising gold deposits. This, together with the want of capital, is the reason the mines have not been opened and worked on an extensive scale. This will follow, however, in due time, for miners can work here all the year round, with comfort as regards the weather, and at the minimum cost of living."

The arrival of an excursion steamer at Sitka is made the occasion of a regular holiday, which is very natural with a people who live in so isolated a place. As the steamer enters the several harbors of the inland passage northward, her presence is announced by a report from the cannon on the forecastle, which awakens a score of sonorous echoes from the rocky cliffs and nearest mountains, also serving to arouse the sleepy natives and put the dealers in curios on the *qui vive*. The few cafés do a thriving business; the nights, never very dark in summer, are turned into day, and hours of revelry prevail. The aboriginal women drive a lively business with their home-made curios, and indiscreet native girls promenade freely with strangers. Peccadilloes are overlooked; no one seems to be held strictly to account. The officials are unusually lenient on such occasions, just as they are in Boston or New York on the Fourth of July.

The immediate environs of Sitka present many rural beauties, including river, forest, and wild flowers, with here and there a rapid, musical cascade. The same species of highly-developed white clover as was seen at Fort Wrangel is a charming feature here, fragrant and lovely, — "Beautiful objects of the wild bees' love." Buttercups and dandelions are twice the size of those which we have in New England. Ferns are in great variety, and the mosses are exquisite in their velvety texture, while tenderly shrouding the fallen and decaying trees they present an endless variety of shades in green. There are over three hundred varieties of wild flowers found on Baranoff Island, and wild berries abound here as among all the islands and on the mainland. The wild raspberry, salmon-berry, and thimbleberry are especially luxuriant and fine in size and flavor. The woods are full of song-birds and of others more gaudy of feather. These are only summer visitors, to be sure, among which the rainbow-tinted humming-bird made his presence obvious. A pleasant walk is finely laid out along the banks of the sparkling Indian River, a swift mountain stream, hedged with thrifty and graceful alders, by which means the citizens have created for themselves a charming and favorite promenade. Along the left bank of this beautiful watercourse are woodland scenes of exquisite rural beauty.

It would be foolish to suggest the idea that

Alaska promises to become eventually a great agricultural country; but it is equally incorrect to say, as did a certain popular writer not long since, that "there is not an acre of farming land in the Territory." There are considerable areas of good arable land now under profitable cultivation in the Sitka district, and large farms, rich in virgin soil, could be had for a mere song, as the saying goes, in desirable localities, by clearing away the timber and draining the land. Some twenty-five milk cows are kept at Sitka; milk is sold at ten cents per quart. Fresh venison is cheap and abundant, and fish of various kinds cost nearly nothing. In the immediate vicinity there are three thousand acres of arable land, much of which is well grassed and covered with white clover. On the foot-hills there is plenty of grass for the sustenance of sheep and goats. Experienced residents told us that wool-growing might be profitably pursued as a business here, and that there was not a month in the year when the animals would absolutely require to be housed. Hay is easily made, and is in abundance at cheap rates. "I have never seen finer potatoes, turnips, cabbages, and garden produce generally, than those grown here," says Governor Swineford in his annual report to the Department at Washington.

There is a great abundance of natural and nutritious grasses in most parts of the country, but especially in the southern islands and the Kodiak group. The great prosperity of Alaska, however, to be looked for in the near future, lies in the energetic development of her coal trade, her fisheries, and her extraordinary mineral wealth. The immense supply of timber, some of which is unsurpassed in its merchantable value, will come into use one or two generations later. The fur-trade, already of gigantic proportions, cannot be judiciously developed beyond its present volume, otherwise the source of supply will gradually become exhausted. It might be quadrupled for a few years, but this would be killing the goose that lays the golden egg. If protected, as our government is striving to do for it to-day, it will continue indefinitely to meet the market demand without glutting or overstocking it. In this connection, and after some inquiry, we cannot refrain from expressing the fear that the legal limit as regards the slaughter of the seals is greatly exceeded. Over three million dollars' worth of canned salmon were exported from Alaska last year. "This Territory can supply the world with salmon, herring, and halibut of the best quality," says Dr. Sheldon Jackson.

Twenty miles south of Sitka, on the same island, there are a number of hot springs, strongly impregnated with iron and sulphur, the sanitary nature of which has been known to the Indians for centuries, and hither they have been in the habit of resorting for the cure of certain physical ills, especially rheumatism, to which they are so liable. Vegetation in the neighborhood of these springs is tropical. The temperature of the water is said to be 155° Fah. At the time of the Russian possession the whites built bath-houses on the spot, and much was made of this sanitarium. But all is now neglected, except that the natives still occasionally resort to the place to enjoy the tonic and recuperating effect of the waters. Anything which will promote cleanliness among the Alaskan tribes must be unquestionably of benefit to them. There are plenty of hot mineral springs all over the various island groups of the Territory, and especially that portion which makes out from the Alaska Peninsula westward towards Asia. The most fatal diseases prevailing among the aborigines after consumption are scrofulous affections; the latter is thought to be aggravated, if not induced, by their almost exclusive fish diet, supplemented by their gross uncleanliness. The Aleuts of the south, the Eskimos of the north, and the natives generally of the coast and the interior sleep and live in such dark, dirty, unventilated quarters, reeking with vile odors, that they cannot fail to poison their blood and thus induce a myriad of ills. As we have said, none of these natives seem to have any intelligent idea of medicine, and they do not possess any herbs, so far as we could learn, which are used for medicinal purposes. If a native is furnished with a prescription after the manner of the whites, he requires at least twice the amount of medicine which it is customary to give to a white man, otherwise the dose will have no apparent effect upon his system. This is a never varying experience which medical men have found repeated among all savage races.

As far as one is able to comprehend the religious convictions of the native Sitkans, other than the few who have gone through the form of professing Christianity, they seem to entertain a sort of animal worship, a reverence for special birds and beasts. Like the Japanese they hold certain animals sacred and will not injure them. It is thus that they have some mystical idea about the bear, which prevents them from willingly hunting that animal. Ravens are nearly as numerous in Sitka as they are in Ceylon, and no one will injure them. They believe that the spirits of the departed occupy the bodies of ravens, hawks, and the like. One is reminded that in the temples of Canton the Chinese keep sacred hogs; the Parsees of Bombay worship fire; the Japanese bow before snakes and foxes, as divine symbols; the pious Hindoo deifies cows and monkeys; so there is abundant precedent to countenance these simple natives of Alaska in their crude worship and superstitions.

Their aboriginal belief is called Shamanism, or the propitiating of evil spirits by acceptable offerings. It is significant that the same faith is participated in by the Siberians, on the other side of Behring Strait. This is no new or original form of religion; it was the faith of the Tartar race before they became disciples of Buddhism.

These aborigines seem to anticipate a state of future happiness, but not one of rewards and punishments. All blessedness in this anticipated eternity is for man; woman, it seems, has no real inheritance in this world or the next! Slavery, vice, and misery would thus appear to be her portion in life, and she expects nothing beyond. This picture is not overdrawn. These natives are now as much a part of our population as are the people who live in Massachusetts or Rhode Island, and our manifest duty is to educate them. The light of reason will soon follow, and like the rising sun will burn away this mist of ignorance and superstition. Schools are the most potent missionaries that can be established among any savage race; reasonable religious convictions will follow as a natural result.

"When the missionary," says W. H. Dall, " will leave the trading-post, strike out into the wilderness, live in the wilderness, live with the Indians, teach them cleanliness first, morality next, and by slow and simple teaching raise their minds above the hunt and the camp, — then, and not until then, they will be able to comprehend the simplest principles of right and wrong." Though these Indians at the populous centres often pretend to yield to the religious teachings of the professional missionaries, still, like the Chinese religious converts, they are pretty sure to return to their idols and superstitions. When the Roman Catholic Bishop from San Francisco came among the natives of Alaska, and offered to baptize their children, the Indians told him that he might baptize them if he would pay them for it!

H. H. Bancroft, in his work upon the native races of the North Pacific, says: " Thick, black clouds, portentous of evil, hang threateningly over the savage during his entire life. Genii murmur in the flowing river, in the rustling branches of the trees are heard the breathings of the gods, goblins dance in the vapory twilight, and demons howl in the darkness. All these things are hostile to man, and must be propitiated by gifts, prayers, and sacrifices; while the religious worship of some of the tribes includes practices frightful in their atrocity."

The Sitkans, like many other tribes, used to burn their dead before the missionaries partially dissuaded them from doing so, but some still adopt cremation as a final and most desirable resort. To one who has seen its universal application in India, there are many strong reasons in its favor. The Alaskan native idea of a hell in another world constituted of ice, it is said, causes them to reason that those buried in the earth may be cold forever after, while those whose bodies are burned will be forever warm and comfortable in the next sphere. After the funeral these aborigines, as we have shown, engage in a genuine "wake," recklessly feasting and drinking to emphasize the importance of the occasion, and to demonstrate their unbounded grief.

The native women occasionally show some taste for music and ability in playing upon the accordion, almost the only instrument found in their possession. A young Indian girl was seen quite alone among the wild flowers just outside the town (Sitka) who had been taught a few pleasing airs, and who surprised us with a well-played strain from a familiar opera. She was a pretty, gypsy-like child of nature, evidently having white blood in her veins, and was not over sixteen years of age. The coarse, scanty clothing could not disguise her handsome form, bright, intelligent face, or hide the depth and splendor of her jet-black luminous eyes. When she discovered us the accordion was quickly thrust behind her, while her downcast eyes expressed mortification at being found alone by the white strangers, playing to the flowers beside the Indian River. She understood English and spoke it fairly well, but hesitated to receive the bright bit of silver offered to her. When we told her that in the East it was the custom to pay those who played to us upon musical instruments out-of-doors, and described the itinerant hand-organist with his monkey, and the brass bands which perambulate city streets, she laughed heartily, thrust the shining silver in her bosom, and held out her hand to greet us cordially. As we turned our steps back towards the town the innocent, winning face of the young girl haunted us with thoughts of hidden possibilities never to be fulfilled.

On the evening before we left Sitka a brass band consisting of twenty-one performers marched down to the wharf from the mission school, in good military order, headed by their teacher as band-master, and serenaded the passengers. The band was composed entirely of native boys, the oldest not over eighteen, not one of whom had ever seen a brass musical instrument two years ago. They performed eight or ten elaborate pieces of composition, not passably well, but admirably, in perfect time, and with real feeling for the music they expressed. It was a surprise to every one on board the Corona to hear such a performance by natives in this isolated spot in the far north. A liberal purse was handed to the teacher to be divided among them.

"Do you know what they will do with this money?" he asked, gratefully.

"Purchase some trifle, each one after his own fancy," we replied.

"No, sir," said the teacher, "they will tell me, every one of them, to purchase some new music with the money, which they can practice and learn to play together."

Their means are of course quite circumscribed, and they have had but little variety afforded them, either in school-books or music. They look upon their musical tuition as a reward for good behavior, and the severest punishment to them is to be deprived of any favorite branch of instruction.

At our final view of Sitka, the quaint capital of Alaska was lying quiet and peacefully at the feet of Vestova, while enshrouded in a voluptuous sheen of afternoon sunlight. A rose-glow rested on everything, beautifying the simplest objects. Lofty, thickly-wooded hills formed the background, while the Greek church and the old castle dominated all the humbler buildings. The waters of the island-dotted bay were as still as an inland lake, and flooded with golden reflections. Now and again an eagle sailed gracefully from one wooded height to another, and the hoarse croak of many ravens, held sacred by the Indians, greeted the ear. A few United States soldiers lounged about their barracks, and a few cannon were arranged upon the broad common. These were light fieldpieces, more for show than for use. Groups of natives clad in bright-colored blankets were seen here and there before their simple dwellings which line the beach. A broad, intensely green plateau forms the centre of the settlement, about which the better houses of the whites are situated. A little to the left, nearer to the hills, is the curiously arranged burial-ground of the aborigines, with a few totem-poles, and many boxes reared above ground in which are deposited the remains of former chiefs. On a slight rise of ground stands the ancient blockhouse, built of logs, from which the Russians once made a desperate fight with the natives. Behind us Mount Edgecombe loomed far up among the clouds, where its apex was half hidden, and in the same direction, not far away, was the open Pacific. It was nearly ten o'clock P. M. before the sun set behind the distant western hills in a blaze of scarlet, yellow, and purple, reflected by soft, butterfly clouds and mountain tops in the east. After that came the luminous moonlight, making a regal glory of the darkness, and flashing in opal gleams from the sea.

While watching the rippling lustre of the water, tremulous with starlight and the languid breath of the night air, one was fain to ask if it was all quite real, if this was not a fancy picture from the land of dreams. Could these be the far-away shores of Alaska? The pathos and tenderness of the scene, the glow, and fire, and throbbing loveliness, were indescribable. Even the few fleecy clouds which sailed between us and the planets seemed as if they came to waft our hymn of praise to Heaven. Is not such surpassing beauty of nature an image of the Infinite One?

CHAPTER XXIII.

The Return Voyage. — Prince of Wales Island. — Peculiar Effects. — Island and Ocean Voyages contrasted. — Labyrinth of Verdant Islands. — Flora of the North. — Political Condition of Alaska. — Return to Victoria. — What Clothing to wear on the Journey North. — City of Vancouver. — Scenes in British Columbia. — Through the Mountain Ranges.

THE return voyage from Sitka by the inland course takes us first through Peril Straits, so named on account of its many submerged rocks and reefs. It is, however, a wonderfully picturesque passage between the two lofty islands of Chicagoff and Baranoff, strewn as it is with impediments to navigation. We pass the Indian village of Kootznahoo, occupied by a tribe of the same name, a people who have always proved to be restless and aggressive, requiring a strong hand to control them. They are peaceable enough now, having been taught some severe lessons by way of discipline. This tribe as a body still adheres to many of the revolting practices of their ancestors, which other Alaskans, who are brought into more intimate relations with the whites, have discarded. They are also said to be more under the influence of their medicine-men, who foster all sorts of vile rites and superstitions, without the prevalence of which their occupation and importance would vanish.

We make our way through the winding channels of the Alexander Archipelago, of which the Prince of Wales Island is one of the largest and most mountainous. It is about a hundred and seventy-five miles long by fifty miles in width; that is to say, it is as large as the State of New Jersey, and in fact contains more square miles. It is mostly covered with dense forests of Alaska cedar, the best of ship-timber. The shores are indented on all sides by fjords extending a considerable distance into the land. Salmon abound in and about this island, which has led to the establishment of several large fish-canning factories, two new ones being added during the past season. The principal native tribe upon the island is known as the Haidas, whose villages are scattered along the coast. The interior of the island is not only uninhabited, but it is unexplored. The shore hamlets are called "rancheries." Each sub-tribe has a special one representing its capital, where the head chiefs live. Their laws seem to be simply a series of conventionalities. The houses of these Haidas are better structures than those of most natives of the Territory, and they surround themselves, as a rule, with more domestic comforts. Woolen blankets appear to be the investment in which all the spare means of the members of this, as well as most other tribes, are placed, and by the number they possess they estimate their wealth. Woolen blankets, in fact, averaging in value from two dollars and a half to three and a half, are the native currency or circulating medium, being received as such when in good condition; and also given out at the trading stations as payment to natives for furs or for any service, unless specie is preferred.

The meandering course of the steamer brings us now before one Indian hamlet and island, and now another; but these villages are very few in number, hours, and even a whole day, being sometimes passed, while on our course, without meeting a solitary canoe or seeing a human being outside the vessel's bulwarks. These islands, as a rule, have no gravelly or sandy beach, but spring abruptly from out the almost bottomless sea, in their proportions ranging from an acre to the size of a European principality.

Now and again we come upon a reach of the shore where it is shelving, and for a mile or more it is bastioned by a course of stones, of such uniform height and even surface as to seem like the work of clever stone-masons. Skilled workers with plummet and line could produce nothing more regular.

In some places, as we quietly glide close in to the shadow of the land, shut in by the morning fog and mist wreaths, the effects are very curious and even startling. It not being possible to see very far up the shrouded cliffs, down whose sides there rush narrow, silvery cascades, with a merry, laughing sound, they often have the appearance of coming directly out of the sky. It seems as though some peak had punctured one of the overcharged clouds, and it was pouring out its liquid contents through the big aperture.

The contrast between a voyage across the open ocean and a sail of two weeks in this inland sea is notable. In the former instance the voyagers find fruitful themes in the vast expanse and fabulous depth of the ocean, the huge monsters and tiny creatures occupying it, the record of the ship's progress, her exact tonnage, and the trade in which she has been engaged since she was launched. Few persons have in themselves sufficient intellectual resources not to become oppressed with *ennui* under the circumstances. Between Puget Sound and Glacier Bay how different is the experience! There is no monotony here; every moment is replete with curious sights, every succeeding hour full of fresh discoveries. The panoramic view is crowded all day long with sky-reaching mountains, scarred by wild convulsions; verdant islands embowered in giant trees; rocky peaks rising from the bottom of the sea to a thousand feet and more above our topmast head; cascades tumbling down precipitous cliffs; Indian hamlets dotted by totem-poles; canoes gliding over the silent surface of the deep channels; inlets crowded with schools of salmon; mammoth glaciers emptying themselves into the sea and forming opaline icebergs sharply reflecting the sun's dazzling rays. There is no time for *ennui* among such scenes as these; the eyes are captivated by the beauty and the variety, while the imagination is constantly stimulated to its utmost capacity.

The flora of this far northern country does not exhibit the wonderful luxuriance and productiveness which captivates us in the tropics, though one gathers some extremely attractive specimens. Neither the flowers, the insects, nor the birds are marked with the brilliancy of color which distinguish those bathed continually in waves of equatorial sunlight. Here, grandeur prevails over beauty; the trees, if not so verdant, excel in size and majesty; the mountains, in height; the rivers, in volume and length; while the glaciers are without comparison in magnitude and power. Here, is simplicity, vastness, magnificence; there, fertility, fragrance, loveliness. Neither in the north nor in the south is there the least infringement upon the great harmonies of Nature; admirable consistency and order exist everywhere, typifying a great, overruling, supreme Intelligence.

We pause for a moment amid the silent tranquillity to sum up our experience while gliding along this beautiful and peaceful inland sea on the return voyage. The author does not hesitate to pronounce Alaska to be one of the most attractive regions in the world for summer tourists. From early June to September the temperature prevailing upon the entire route is equable, the thermometer ranging all the while between sixty and seventy degrees Fah. The progress of the steamer always creates a gentle and agreeable breeze, which renders warm clothing desirable, especially at early morning and in the evening, though these are periods not so distinctly defined as with us in New England. An overcoat is rarely rendered necessary or desirable. If the mosquitoes are troublesome at certain places on shore, in marshy regions, they are never so on the water, as the breeze inevitably drives such insects away. Let us say especially there is no other such inviting resort for pleasure yachts as this inland, island-dotted sea of Alaska. If the fogs put in an appearance sometimes in the morning, they are after a while burned away by the warmth of the sun. Local rains on shore are to be occasionally endured, but they are no great drawback to observation and brief excursions. At Sitka, Wrangel, and Juneau several showers may occur during the day, with intervals of bright and cloudless skies between. We have witnessed seven copious, well-sustained showers of rain on a May forenoon in Chicago, the intervals sandwiched with sunshine of gorgeous clearness and warmth. Why pretend that Alaska is exceptional in this respect? The weather is not perfect, according to our estimate, anywhere. Finally the extended trip upon the boat was found to cover a little over two thousand miles in all, and was with us one of continuous pleasure, enlivened by as bright and cheerful weather as one experiences on an average elsewhere, winding among an immense archipelago of mountains, emerald islands, and land-locked bays, through narrow channels dominated by precipitous cliffs, and crossing broad, lake-like expanses as placid as the serene blue overhanging all.

No other government on the globe, in this nineteenth century, would permit so large and important a portion of its territory to remain unexplored. Congress should send at once a thoroughly equipped scientific expedition, competent to report minutely upon the geology, fauna, flora, and geography of this immense division of the country. It is more than an oversight, it is a gross blunder, not to do this without further delay. If our own pen-pictures of this neglected Territory shall incite to the fulfillment of such an act of official duty, these pages will have served at least one important purpose.

"With a comparatively mild climate," says C. E. S. Wood, in an account of a visit to Alaska, printed in the "Century Magazine," "with most valuable shipbuilding timber covering the islands, with splendid harbors, with inexhaustible fisheries, with an abundance of coal, with copper, lead, silver, and gold awaiting the prospector, it is surprising that an industrious, shipbuilding, fishing colony from New England or other States has not established itself in Alaska."

The political condition of Alaska is anything but creditable to our country. It has little more than the shadow of a civil government, and is entirely without any land laws by which a resident can secure a title to the soil upon which he builds his house. The act of Congress dated May 7, 1884, providing an apology for a civil government, was not passed until twenty years after the Territory had been acquired. As a consequence the material progress of the country and its inviting possibilities remain undeveloped. With the extension of the United States local laws to this section, immigration would be at once promoted and various industries established. "Why we are so neglected is incomprehensible," said a resident of Sitka. "All we ask is the same advantages enjoyed by the citizens of the other Territories of the United States." It is certainly to be hoped that Congress will give early attention to this important matter, for Alaska is destined to become one of our most valuable possessions. We shall be excused for making use of so strong an expression, but it is only too true that her interests have been persistently and shamefully neglected by the law-makers at Washington.

"Like the dog in the manger," says Miss Kate Field, "Congress will do nothing for Alaska, nor will it permit Alaska to do anything for herself locally, or at Washington through a delegate. Yet, in 1890, two islands of this despised and neglected province will have paid into the United States Treasury $6,340,000, — within one million of Alaska's entire purchase!"

The present comparative isolation of Alaska will not be of long duration; not only are the facilities for reaching the Territory being annually increased from the east, but it is being also rapidly approached in this respect from the west. The Russian government is building a railroad in almost a straight line from Moscow to Behring Sea, which it is confidently believed will be completed within five years. Direct communication will thus be established between St. Petersburg and the Russian Pacific ports, through Siberia, whose most easterly point is less than forty miles from the soil of Alaska.

After sailing four or five days southward, bearing always slightly to the east, through a wilderness of islands and along the mountain-fringed coast of the mainland, the ship comes upon the open sea, and the passengers realize for a short time the effect of the Pacific Ocean swell. The sensitiveness of some people to its influence is as remarkable as the stolid indifference of others. Here, where the Japanese Current meets the cold air from off the coast, fogs are very liable to prevail, though it was not so in the writer's case. We are now in comparatively open navigation and can lay our course without fear. Soon Queen Charlotte's Sound is entered, and for a day and a half the steamer again skirts the picturesque shore of Vancouver, whose features are reproduced in the deep, quiet waters with marvelous distinctness, until finally we are once more landed at Victoria, the capital of British Columbia.

We are frequently asked since our return what clothing and other articles one should take, with which to make the inland voyage through Alaskan waters. This is easily answered.

As the rainfall is frequent be sure to have a good stout umbrella. Ladies would do well to take a gossamer waterproof and gentlemen a mackintosh. Heavy shoes, that is with double soles, and a light overcoat should be provided. There is no occasion for full dress,— court dress, on this route, swallow-tails are so much needless baggage. Ladies' skirts should be short so they will not draggle on the wet deck of the steamer, or in walking through the damp grass, or over the surface of a glacier. In the latter instance gentlemen generally carry portable spikes that can be screwed on to the bottom of the shoes, and a staff cane with a stout ferule. When a party is formed to ascend a glacier a small hatchet and small rope should always be taken by some one of their number. In case of an accident these often become of great importance. There need not be any accident, however, if ordinary prudence is observed.

A large and well-appointed steamer named the Islander, which plies regularly on this route, takes one across the island-sprinkled Gulf of Georgia in six or seven hours from Victoria to Vancouver on the mainland. This is the terminus of the Canadian Pacific Railway, situated a short distance from the mouth of the Fraser River. From here the homeward course is almost due east through British Columbia, Alberta, Assiniboia, Manitoba, Ontario, and Quebec to Montreal, thence southeast to Boston.

So late as 1886 the present site of Vancouver was covered with a dense forest of Douglass pines, cedar and spruce trees. The Canadian Pacific Railway was completed to Vancouver in May, 1887, when the first through train arrived from Montreal. The youthful city is well situated for commercial purposes on what is called Burrard Inlet. It has extensive wharves, substantial warehouses, and very good hotel accommodations. Well-arranged public water-works bring the needful domestic supply in pure and healthful condition from the neighboring hills. The surrounding scenery is strikingly bold, embracing the Cascade Range in the north, the mountains of Vancouver Island across the water in the west, and the Olympian Range in the south, while the great snowy head of Mount Baker rears itself skyward as the main feature in the southeast. The steamer which brings us here from Victoria passes through a beautiful archipelago of peaceful islands, verdant and wooded to the very brink. The busy population of this infant city number between thirteen and fourteen thousand, and the place is growing rapidly. It is lighted by both gas and electricity. Forty substantial edifices for business and dwelling purposes are in course of erection at this writing. There are steamers which sail regularly from here for Japan, China, and San Francisco. As it is in the midst of what may be called a wild country, there is excellent hunting near at hand and large game is abundant. Many sportsmen, especially from England, make their headquarters here while devoting themselves to hunting for a large part of the summer season. Four large English sloops of war were observed in the harbor at the time of the writer's visit, together with a couple of torpedo boats bearing the same flag, destined for Behring Sea, to "emphasize" the British side of the Alaska fishery question as between our government and that of Great Britain.

As one stands on the shore the harbor presents a picture of great variety and interest, comprising men-of-war boats pulled by disciplined crews; canoes, paddled by Indian squaws wrapped in high-colored blankets; boats loaded with valuable furs and propelled by aboriginal hunters; here a raft of timber, and there a steam ferry-boat. Just in shore there is passing as we watch the scene a native canoe carrying a sail made of bark-matting, brown and dingy, steered with a paddle by an aged, withered, white-haired Indian, while in the prow is a four or five year old native boy, trailing his hands idly in the water over the side of the tiny craft. A striking picture of the voyage of life: thoughtless, happy, vigorous youth at the prow, with weary age and experience awaiting the end at the stern. A couple of large steamers close at hand are getting under way loaded with preserved fish, put up at the canneries near by; one is bound for Australia, the other for England, by way of Cape Horn.

Vancouver has many edifices of brick and stone, with good churches and several schools; some of the private residences being remarkable for their complete architectural character in so new a city as this which forms the terminus of the Canadian Pacific Railway.

The principal part of the city occupies a peninsula, bounded north by the waters of Burrard Inlet, south by a small indentation called False Creek, and west by English Bay. The city is fast extending beyond these limits, both east and south.

The peninsula rises gradually to an altitude of two hundred feet, more or less, affording the means of perfect drainage for the new city, which is laid out on a grand scale. A tramway, embracing the several suburbs, is in course of construction, the motor for which will be electricity.

We take the cars at Vancouver for our long journey homeward over the Canadian Pacific Railway, through the British Dominion to the Atlantic coast, indulging in a last admiring view of the grand elevation known as Mount Baker, which in these closing days of July is a mass of snow two thousand feet from its summit. Upon starting our attention is first drawn to the gigantic trees, big sawmills, immense piles of lumber, and extensive brick-yards in the environs of the city. Small villages are passed, straggling farms, Indian camps, mining lodges, and Chinese "hives," where these people congregate after working all day at placer mining, and gamble half the night, sacrificing their laboriously acquired means. The grand winding valley of the Fraser River — a watercourse as large as the Ohio — is followed for over two hundred miles in a northeasterly direction, affording glimpses of most charming and vivid scenery, leading through cañons fully equaling in grandeur of form and beauty of detail anything of the sort in Colorado.

Now and again groups of Indians are seen preparing the salmon they have caught for winter use. The fish are split and stretched flat by wooden braces, then hung in long pink lines upon low frames of wood. They use no salt in this curing process, but simply dry the fish by atmospheric exposure, and succeed very well in thus preserving it. Dried salmon forms the principal staple of food for this people in the long Canadian winters. These natives, as in our own instance, are subsidized by the Dominion; that is, they are placed upon reservations and receive a certain amount of money and rations annually from the government. Light green patches of raspberries are passed here and there, where children are gathering the ripe fruit in abundance, the bright color about their mouths betraying how abundantly they have feasted while thus engaged. It was a pleasant picture to gaze upon under the pearly blue sky, where we were surrounded with the fragrant odor of pine and spruce, and the ceaseless music of hurrying waters.

At times the river rushes through deep rocky ravines, and at others expands into broad shallows with glittering sand bars, on which eager groups of miners are seen washing for gold. We cross a deep, cavernous gorge of the river on a graceful steel bridge, which, though doubtless of ample strength, yet seems of spider-web proportions, then plunge into a dark tunnel to emerge directly amid scenery of the wildest nature, set with huge bowlders and noisy with boiling flumes and roaring cascades, where color, splendor, and inspiration greet us at each turn, while every object is softened by the pale afternoon sunlight.

By and by we pass up the valley of the Thomson River, a tributary of the Fraser, finding ourselves presently in what is called the Gold, or Columbian, range of mountains, a grand snow-clad series of hills. Our route through them for nearly fifty miles is in the form of a deep, narrow pass between vertical cliffs, forming land channels similar to the water-ways which we have lately left behind us in the Alexander Archipelago.

At the small stations boys and girls board the cars with tiny baskets of luscious blackberries and ripe raspberries for sale, soon disposing of them to the passengers. These are picked within a dozen rods of the railway track, where they are seen in great abundance. Wild flowers beautify the roadway, among which the most attractive are the golden-rod, the bright pink fire-weed, the towering and graceful spirea, the wild musk with its large bell-shaped scarlet flower, the fragrant tansy, with snow-ball clusters of white, and big patches of the tiny wild sunflower, its petals in deepest yellow, while among the lily-pads dotting the pools of water, orange-hued lilies are in full and gorgeous bloom.

The scenery is strictly Alpine, but constantly varies as our point of view changes, and we thread miles upon miles of snow-sheds. Heavy veils of mist fringe the mountain-tops, and the tall peaks are wrapped in winding-sheets of perpetual snow. The rugged scenery is fine, but finer is yet to come. Still climbing upwards, we are presently in the Selkirks, threading tunnels, dark gorges, sombre cañons, and narrow passes to the summit of this remarkable range, forced onward by two powerful engines, one in the rear the other in front of the train.

At a point known as Albert Cañon the railway runs along the brink of several dark fissures in the solid rock, three hundred feet deep, through which rushes the turbulent waters of the Illicilliwaet River ("Raging Waters"). Here the cars are stopped for a few moments that the passengers may the better observe the boiling flumes of angry waters, flecked with patches of foam, and compressed within granite walls scarcely twenty feet apart.

In approaching Glacier House station, at a certain point the train ascends six hundred feet in a distance of two miles. This is accomplished by a zigzag course, utilizing two ravines which are favorably situated for the purpose; the consummation is a grand triumph of engineering skill. While passing through this winding course we are serenaded by a chorus of dancing rapids, foaming cataracts, and rushing cascades. Here the torrents and waterfalls are innumerable, first on one side then on the other of our slowly-climbing train, and finally on both the right and the left, gleaming with bright prismatic rays while moving with tremendous impetus. Sir Donald, the highest peak of the Selkirk Range, shaped like an acute pyramid, now comes into view, rising to eleven thousand feet above the level of the sea, and piercing the blue zenith with its inaccessible summit. It is named after one of the most active promoters of this transcontinental railway. Sir Donald sends down from its immense snow-fields a ponderous glacier half a mile wide and eight miles long, presenting most of the characteristics of such frozen rivers, though lacking the grand effect of those so lately seen in Alaska, where they join the ocean in partially congealed form, thus producing thousands of icebergs. This Donald glacier is nevertheless equal to the average of European ones. The mountain has never yet been ascended. We were told that a thousand dollars and a free pass over the railway for life await the successful mountain-climber who reaches the summit.

In making our way through Beaver Cañon and Stony Creek Cañon, the highest timber railway bridge ever constructed is passed, three hundred feet high and four hundred and fifty long, supported by direct uprights. Safe enough, perhaps, but one breathes freer and deeper when it is passed.

It would seem as though mosquitoes could hardly thrive at such an altitude, but their number here is myriad, and their vicious activity at Glacier House station beggars description.

CHAPTER XXIV.

In the Heart of the Rocky Mountains. — Struggle in a Thunder-Storm. — Grand Scenery. — Snow-Capped Mountains and Glaciers. — Banff Hot Springs. — The Canadian Park. — Eastern Gate of the Rockies. — Calgary. — Natural Gas. — Cree and Blackfeet Indians. — Regina. — Farming on a Big Scale. — Port Arthur. — North Side of Lake Superior. — A Midsummer Night's Dream.

ROGERS' PASS, at an altitude of four thousand two hundred and seventy-five feet above the sea, is situated between two ranges of snow-clad peaks, whence a dozen glaciers may be seen in various directions, frigid and ponderous.

As we came through this remarkable pass, in the afternoon, dark clouds rapidly spread themselves over the sky, reinforced by others more dense and threatening, engulfing us suddenly in darkness. Then the artillery of the heavens rang out in such deafening reports as to stifle all attempts at speech. The discharges and echoes among the gloomy gulches and tall peaks mingled so rapidly that it was impossible to separate cause and effect. The rain was like a cloud-burst. The sharp flashes of lightning were so incessant and blinding that one sat with closed eyes and bated breath. The great locomotive could barely make way on the steep up-grade, the wheels having so much less hold upon the track when thus submerged. Passengers looked into each other's pale faces in fear and amazement. Still the slow, regular *throb, throb*, of the iron horse was heard through the din of the thunder and the roar of rushing waters. We did move forward, — barely moved. To stop would be destruction; backward impetus would instantly follow, and no brakes are powerful enough to stop the train from a dash downward towards the plain if once it started in that direction. But stay. Soon there came a faint glimmer of light from out of the sky, gradually this increased, the dark pall of the heavens was slowly removed, and the afternoon sun burst forth with soft, ineffable beauty. The thunder sounded farther and farther away, the echoes ceased, and the *throb, throb* of the ponderous engine steadily held the long train and forced the great load onward.

At Field station, in the heart of the Rocky Mountains, we begin an ascent of twelve hundred and fifty feet with two powerful engines, where the roadway is cut out of the sides of nearly perpendicular cliffs to which it seems to cling with iron grasp, overhanging the roaring torrent of the Kicking Horse River, which flows at a fabulous depth below. Here we cross now and again trestle bridges, three hundred feet above some frightful gorge, or pass over a viaduct of great span. The highest point of the road is reached at fifty-three hundred feet above the level of the sea, or say just one vertical mile. This extreme elevation is about five hundred miles from Vancouver.

The scenery at this point is grand beyond description, thrilling the whole nervous system while we gaze at it and vainly strive to comprehend its vastness. The very excess of emotion makes one dumb. The most experienced traveler watches the changing scene with a vivid interest. So wild, so comprehensive, and so startling a natural panorama is rarely met with in any land. A longing comes over the observer to divide the ecstasy of the moment with the loved ones left behind. No joy is complete which is not shared; it is no hermit quality, but was born a twin. Mountains, valleys, glacier-bound peaks, domes, spires, and snow-capped pyramids are seen in all directions, brought out in minute detail by the singular clearness of the atmosphere. Tall forests are spread out far, far below our feet, the mammoth trees looking no larger than pen handles, while the river winds like a broad silver belt through the green sward of the valley. Thus the Canadian Pacific Railway passes for hundreds of miles along glacial streams in full sight of the frozen rivers which feed them.

By and by we come in view of Castle Mountain, five thousand feet in height, which, with a little help of the imagination, becomes a giant's keep, turreted, bastioned, and battlemented. At another point of view it presents a remarkable resemblance to the grand Indian Temple of Tanjore. A short distance farther and we reach Banff, where a couple of days were most agreeably passed by the author. The railway station here is in the midst of sky-piercing heights, whose first impression upon the traveler is both solemn and lonely. To the northward stands Cascade Mountain, nearly ten thousand feet in height; eastward is Mount Inglismaldie, beyond which looms up the sharp cone of Mount Peechee, reaching more than ten thousand feet into the blue ether. Close at hand rises the thickly wooded ridge of Squaw Mountain, in whose shadow lie the beautiful Vermilion Lakes, the home of myriads of wild geese and ducks. Other mountains are in view, but in the memorable tableau which we recall the grand peaks we have mentioned are the most prominent.

This is the station for the Rocky Mountain Park, the altitude being forty-five hundred feet above the sea. At this point the Canadian government has established a national reservation after the plan of our Yellowstone Park, between which and this place lies five hundred miles of the wildest sort of country. There is no comparison between the two parks, either in size, importance, or natural wonders. This reservation is twenty-six miles long by ten in width, embracing portions of three rivers, with two considerable lakes, cascades, and waterfalls. The scenery could not be otherwise than bold, being in the midst of such a mountain range and surrounded by such monarch elevations. Money is to be freely expended in making good paths, together with convenient avenues and bridges.

The Pacific Railway Hotel at Banff is a large,

admirably situated, and picturesque establishment, designed to accommodate from two to three hundred guests at a time, and is especially patronized by Canadian bridal parties. The view from it is superb, commanding the winding course of the Bow River and valley for miles, with the many adjacent mountains. The river pours swiftly down from its sources among the snow fields, and plunges seventy feet over rock and precipice close beside the hotel, passing almost beneath our feet as we stand upon the broad piazza, gazing in admiration at the grand scenic carnival, and listening to the thrilling anthem of the rushing waters, while breathing the soft aroma of the Douglas pine and cedar forests which cover the surrounding slopes. The region in proximity to the hotel will give the lover of fishing ample sport. Trout of large size abound in Devil's Lake near at hand. A guest brought in forty pounds of this gamey fish, caught in two hours' time in the lake, while the author was at Banff. Wild sheep and mountain goats abound in the neighboring hills, while bears are more numerous than is desirable. Wildcats, mountain lions, deer, and caribou are also frequently shot by the hunters. The restriction as to use of firearms which is established in the Yellowstone Park does not apply in this region. Sportsmen roam where they please and freely hunt the wild animals which roam in this section of the country. Good roads and bridle paths take one in all directions among some of the finest scenery of the Rocky Mountains, where we watch the morning sun dispel the mist which floats upward and away, disclosing the snow-decked peaks in their virgin whiteness blushing roseate tints at the ardor of the sun.

This is called the eastern gateway to the Rocky Mountains, through which the grand Bow River flows on its diversified journey of fifteen hundred miles to Hudson Bay.

There are extensive hot springs on the eastern slope of what is known as the Sulphur Range, some six thousand feet above the sea level. They are at different elevations, and have good bathing-houses erected over them, in charge of courteous attendants. One of the springs is inside of a dome-roofed cave, which is a favorite resort of visitors to Banff. The medicinal character of these springs is considered so important that an iron pipe two miles in length conducts their heated waters for use at the hotel, the normal temperature being sustained by metallic coils of superheated steam. It rains much and often in this region. The weeping clouds make one feel rather gloomy, purely out of sympathy for their ceaseless tears, but when the sun finally asserts his power and lifts the misty veil, then come forth in bold contrast silvery, sparkling, sky-reaching mountains, covered with their frosty mantles, together with richly wooded valleys and river-threaded cañons, opening views of unrivaled sublimity and grandeur.

At Anthracite, five hundred and seventy miles from Vancouver, we are forty-three hundred and fifty feet above the sea. Here are the remarkable coal mines located in the Fairholme Range, a true anthracite of excellent quality and of great importance to the railway. The pass through which the road takes us is four miles wide, great masses of serrated rocks rising on either side, back of which mountains tower above each other as far as the eye can reach, forming long vistas of lofty elevations so numerous as not to bear individual names.

At Calgary, about a hundred miles farther eastward, we are still thirty-four hundred feet above the sea. This is a particularly handsome and thriving young town, scarcely four years old, but containing three thousand inhabitants. It is pleasantly situated on a hill-girt plateau, in full view of the jagged peaks of the Rockies, thirty or forty miles away, and which, as we look back upon them, form a vast blue and white crescent extending around the western horizon. Two placid rivers, the Bow and Elbow, wind through the broad green valley, adding a charming feature as they mingle with the tall waving grass. Here cattle and sheep ranches abound, extending westward to the very foot-hills of the great mountain range, and stretching far away to the southward a hundred and fifty miles to the United States boundary line. We were told that the cattle and horses ranging over this space would aggregate two hundred thousand head.

As we passed through the Province of Alberta at night, occasionally jets of flaming natural gas, which finds vent through the soil from reservoirs located at unknown depths, were burning brightly to light us on the way. This gas, so liberally supplied by nature free of cost, is utilized to create a motive power at Langevin, where it pumps water for the use of the railway. Representatives of the aboriginal Cree and Blackfeet tribes form picturesque groups along the railway line, composed of barbarous, uncleanly looking squaws and bucks, the latter only kept from the warpath by the presence of the efficient mounted police.

The contrast presented in emerging from the mountain ranges on to the level country is very remarkable. For hundreds of miles we pass through an almost uninhabited, treeless country, a long, long reach of prairie as boundless as the sea, and where no more of human life is seen than on the ocean. There are no hills, scarcely any undulations; the sun rises apparently out of the ground in the early gray of the morning, and sets in the endless level of the prairie at night. Small stations, twenty or thirty miles apart, have been built by the Canadian Pacific Railway Company, consisting of a dwelling-house and a water-tank for the necessary supply of its engines, but the line is thus characterized through a thousand miles, where there is no way travel, and no local business, outside of its own necessities. The inference is plain that it crosses this distance at extraordinary expense, which must be supported by the terminal business on the Pacific and Atlantic ends of the road.

The Cree and Blackfeet tribes are said to have no religion and few superstitions, being a restless, dangerous race, ranking very low in point of intelligence, even as savages. The efforts of the missionaries, we were told, have entirely failed to civilize or even permanently to improve the condition of the two tribes we have named. The women are hideously ugly, smeared with vermilion, and weighed down with cheap brass rings and bracelets of the same metal. The one article of sale offered to the traveler by these tribes is the polished horns of the buffalo, picked up upon the vast prairies of this region where they have been bleaching for many years. These are colored black by some process, and when highly polished are mounted in pairs, as they are placed by nature on the animal's head.

At Regina, eleven hundred miles from Vancouver, we are still two thousand feet above the sea. This is the capital of the Province of Assiniboia, situated in the centre of an almost boundless plain. Here are the headquarters of the Northwestern Mounted Police, a very necessary military organization of a thousand men, distributed over this region to look after the Indians, who are ever ready to commit depredations when they feel they can do so with impunity, and also to preserve good order generally among the several frontier communities. It was at Regina that Louis Riel, the principal promoter of the late rebellion against the Dominion government, was tried and hanged not long since. It is called here the "half-breed rebellion." Over the far-reaching, trackless, arid prairies, as lonely as an Egyptian desert, the cloud effects towards the day's close are noticeably very fine, while the twilight lingers to the very verge of night. At times we pass through a broad tract of land ten miles or more square, from which a whole forest has been swept by conflagration, probably started by an unfortunate spark from a passing locomotive, or, quite as likely, by the carelessness of some camping party of sportsmen. These large spaces, which would otherwise be intensely dreary, are already carpeted with a fresh green undergrowth, with which nature always hastens to obliterate the devastation caused by the ruthless flames.

As our train stopped briefly at Regina a group of mounted Blackfeet Indians dashed across the prairie and drew up near the station. A wild, weird score of semi-savages, very picturesque in their garments of many colors and their decorations of quills, beads, and feathers, with a scalp hanging from the waist here and there among them. Their long, unkempt black hair flowed all about their necks and features, which were more or less besmeared with vermilion. Their leggings of deer-hide were fringed on the outer side, and their leather moccasins were lashed with deerskin thongs up the ankles. Some had stirrups, but most of them had none, their limbs hanging free and a blanket serving for a saddle. Their little wiry ponies were under complete control, and the riders were good horsemen. It seemed to be some gala occasion with these Blackfeet, but of what purport it was impossible to discover. They were evidently under a certain degree of discipline, for at a sharp, sudden command from one of their number they all dismounted together and stood with one arm over their horses' necks like so many stone statues. At that moment a lady passenger in our car aimed her "kodak" at them, and, presto! they were photographed in the twinkling of an eye, which, considering their aversion to the process, was quite an achievement on the lady's part. These Indians are now peaceable enough, and no one fears to go among them, but we are inclined to think, with "Buffalo Bill," that they will make one more desperate fight, in both Canada and the States, before they finally give up the struggle with the white man.

Forty miles eastward from Regina we come to Indian Head, which is about three hundred miles west of Winnipeg, where the road passes through the famous Bell Farm, an extremely interesting and successful agricultural enterprise. It is managed by Major Bell, an ex-army officer of marked executive ability, and covers an area measuring one hundred square miles, being probably the largest arable farm in the world. Major Bell carries on the business for an incorporated company, and devotes the rich prairie loam, of which the soil is composed, mostly to the raising of wheat, employing in the various departments over two hundred men. The announced object of the company is first to bring the whole of the land under good cultivation, at the rate of five thousand acres or more annually, and when this is accomplished to divide the whole into two hundred and fifty farms to be sold to the employees, each provided with suitable dwelling-houses and buildings, all to be paid for by the purchasers in easy annual installments; a most beneficial purpose, and if it is fairly and honorably carried out it will be one which is deserving of all praise. It must inevitably build up a responsible and self-respecting community, by uniting proprietorship and domestic relations of the most desirable character, connected with steady and remunerative occupation.

The country lying between Indian Head and Winnipeg is mostly of a prairie character, rich in agricultural resources but of no special interest otherwise. Winnipeg, the capital of Manitoba, is very nearly midway between the Atlantic and Pacific oceans. It has some twenty-three thousand inhabitants, who live upon a site which was fifteen years ago known as Fort Garry, only a fur-trading station, said to be hundreds of miles from anywhere. To-day it has long, broad streets of public buildings, fine dwelling-houses, hotels, stores, banks, and theatres, besides large manufactories in various branches of trade. It is the Chicago of Canada. Situated where the forests end and the prairies begin, with river navigation in all directions, and with railways radiating from it towards all points of the compass, everything tends to make Winnipeg the commercial metropolis of the British possessions in the Northwest.

Main Street, Winnipeg, is a fine boulevard one hundred feet wide and two miles long, lined from end to end with attractive buildings. One practice observed here recalled the native city of Jeypoor, India, namely, the driving of single oxen to harness between the shafts of light carts, the animal being guided by rope reins attached to the horns.

From Winnipeg to Port Arthur, which is beautifully situated on the north side of Lake Superior, the route is through a country characterized by a maze of forests, lakes, and rivers; a region more than half wilderness. Few evidences of civilization are found hereabouts; the primeval forest is full of game, the streams abound in fish, and the ponds are covered with wild fowl. Occasionally a group of Indian wigwams is seen, or a lone native Chippeway paddling his birch canoe. Now and again a hunter's camp is passed, whose occupants come down to the railway to see the passing train, and who eagerly seize upon any current newspaper which thoughtful passengers toss to them from the car windows, a courtesy they gratefully acknowledge cap in hand.

Port Arthur, just one thousand miles from Montreal, is admirably situated on Thunder Bay, where the view is striking and beautiful, overlooked by the bold headland known as Thunder Cape, which rises fourteen hundred feet above the surface of the lake. Just upon the edge of the horizon is seen Silver Islet, which has heretofore proven to be one of the richest silver mines known to our times; but the mine is now hopelessly submerged, its tunnels and shafts flooded beyond relief by the waters of Lake Superior. These broad waters are dotted with white sails, and streaked with the long black lines of smoke trailing after huge steamers.

From here, for more than one hundred miles, the sharp curves of the great lake on its northern shore are closely followed by the Canadian Pacific Railway, and here the engineer's skill has been wonderfully displayed in surmounting apparent impossibilities. We were told that it cost more per mile to build this portion of the road than it did to lay the rails through an equal distance in the difficult passes of the Rocky Mountains. The roadway is sometimes cut through solid rock, and sometimes an abrupt cliff is tunneled, from whence we emerge to leap across a deep ravine upon a wooden trestle of frightful curve and great elevation. And so we rush onward through unbroken forests and scenery of wildest aspect among barren rocks, scorched trees, and dense thickets of scrub on our homeward way.

Having thus brought the patient reader so nearly back to the starting-point, and among scenes so familiar, we leave him to finish the journey to Boston by way of Ottawa and Montreal.

The distance traveled in making this round trip to Alaska and back, over the course pursued by the author, is something over ten thousand miles, but when successfully consummated it is difficult to realize that such a long route has been passed over. Great are the modern facilities for travel, and great are the inducements. It is the only royal road to learning, the kindergarten of ripened intelligence, so to speak. We recall nothing of the fatigue or the inevitable mishaps of the journey. It is the charming experiences alone which become indelible. We behold again the many populous cities through which the route has taken us, and see once more in imagination the active villages, peculiar races of people, grazing herds, rushing cascades, sombre gorges, mysterious geysers, snowy mountain ranges, uncouth totem-poles, myriads of icebergs, and mammoth glaciers. To look back upon the experiences of the journey as a whole is like recalling a midsummer night's dream, replete with delightful scenery and crowded with wonderful phenomena.

www.ingramcontent.com/pod-product-compliance
Lightning Source LLC
Chambersburg PA
CBHW020229240426
43672CB00006B/466